NANOCOSM

CONTENTS

PREFACE

THIS BOOK IS not a compendium, nor do I present it as an exhaustive or definitive survey. Like all my science writing, it is a highly subjective take on an area that fascinates me: a book-length column. While ideas retain top billing, I do not neglect the places and personalities that reveal these abstract concepts as the lively human things they are. Moreover, I frequently stray from the beaten path of Nobel nominees and huge, well-endowed universities to look into the corners and alleyways of R&D. Gold is where you find it: Money and fame do not necessarily correlate with good ideas.

Nanotechnology is also a young discipline, and like every youngster can be prey to shills and charlatans. I hope to smoke these out when I encounter them, or at least to get my readers asking tough questions about some areas—particularly molecular manufacturing—that despite vast amounts of self-promotion do not yet, and may never, exist.

My aim in this book is twofold. First, to entertain—not with frivolity or unreality, but with that most riveting of things, the scientific and technical frontier. My second aim is to give that unsung hero of the world economy, the venture capitalist, a thorough briefing in the science and technology emerging from the nanocosm. VCs know management and money; but as the dot-com bubble proved, they are too often babes in the wood when it comes to technical matters. One could consider this book, whose reader need understand math and science only at a high-school level, as a gift to the business people whose unending efforts continue to create wealth throughout the globe.

William Illsey Atkinson
North Vancouver, Canada
March 13, 2003

NANOCOSM

LOWER, SLOWER, SMALLER
TOWARD A WORKABLE NANOTECHNOLOGY

*The wise man looks into space and does not regard
the large as too large nor the small as too small,
for he knows there are no limits to dimensions.*
— The Upanishads

THE MOST AMAZING thing about nature is her inexhaustible variety. Scientists, technologists, and theologians speak about "nature" or "the world" as if it were a unit. But there are limitless worlds and infinite natures. Every human brain, its loves and hates and memories, has been correctly described as a three-pound universe. The Pennsylvanian biologist Loren Eiseley, as great an essayist as he was a scientist, invented the concept of "weasel space" to describe the world a nonhuman mammal sees. In weasel space humans, to our own self-centered minds the pinnacles of creation, don't matter at all.

Merely varying your dimensional scale creates new worlds. Karl Marx is justly discredited as a social philosopher, but one of his points was incontrovertible: Quantitative difference creates qualitative difference. In other words, scale matters—change the number and you change the thing. That premise underlies this book.

For example, our usual human view, looking out from the surface of a rocky planet, differs from what we see from the orbit of the moon. All cosmonauts and astronauts agree the most conspicuous thing about viewing earth from space is the invisibility of national borders. The earth appears as a single entity, bearing humans not as a hodgepodge of warring clans but (in Carl Sandburg's wonderful phrase) "travelers together on this bright blue ball in nothingness."

Reality alters further at still larger scales. Imagine the world of the Milky Way—a glowing Catharine-wheel of stars that light takes tens of millennia to plod across, haloed with ancient globular clusters and wrapped around a huge star-swallowing black hole. A common computer screen saver called Star Fields projects white dots onto a black background. The dots emerge from screen center and accelerate out toward the edges; looking at them, you feel you're gazing out a starship's forward viewport. If Star Fields were real, you would have to watch that screen for eighty years just to fly by every star in the Milky Way. And there are, by best estimate, as many galaxies in the universe as there are stars in our mid-sized galaxy, a trillion or so. That's 1,000,000,000,000,000,000,000,000 stars altogether; eighty trillion years of Star Fields. Scale matters.

As a writer specializing in technology and science, I've always loved Douglas Adams's introduction to *The Hitch Hiker's Guide to the Galaxy*. "Listen," it goes. "Space is big. Really, really big. You have no idea how mind-bogglingly, stupendously big space is. I mean, you may think it's a long way down the street to the chemist's (i.e., drugstore) but that's nothing to space. . . ."

This strikes the perfect note. As you read it, you realize that what at first seemed silly is instead deeply earnest: a near-hysterical attempt to convey what words cannot. The numbers are just too large.

You may not realize, however, that even this galactic macrocosm pales when we reverse the direction of our imaginative voyage. All we need to do is turn inward rather than outward, and peer with ever-higher magnification into the world of the small. The traditional Chinese conception of our everyday world is exact. It is *ch'ung k'uo*, the middle kingdom. It, and we its inhabitants, are poised delicately between the unimaginably immense and the unimaginably minute.

To illustrate: The number of stars in our galaxy is *less than half* the number of cells in an adult human body. "I am large," Walt Whitman sang, "I contain multitudes." So do we all. You, reader, exist in finer detail than all the stars and nebulae in the Local Galactic Group. Your structure and function are more complicated; at any given instant your body hosts a greater range of chemical events than all the visible stars. Shakespeare's Hamlet said it best about humanity: *What a piece of work!* The poorest, most broken-down human represents a product, a process, and an achievement that are beyond our comprehension, let alone our imitation.

Below the middle kingdom, which we measure in yards or meters (40 inches), a series of subworlds exists in nested shells. Each subworld embodies an alternate reality. A scale of millimeters brings us to the world of the insects. These mobile computers (the insects) are perfectly adapted to the intricate ecosystems they occupy, from lawns and trees to walls and mattresses. Drop down a notch and you enter the world of the micrometer or *micron*, a unit of length that is one thousandth of a millimeter. This subworld is, literally, the microcosm. It is the world of the cell—autonomous units such as amoebas and zooplankton, as well as specialized "social" populations that make up skin, bones, and brain.

Below the microcosm comes creation on the scale of the *nanometer*, one millionth of a millimeter. I call this the nanocosm. It is a finely detailed, completely structured cosmos, or organized universe, that exists around and within us. All that is—microbes, humans, planets, stars, totality—is built up from the nanocosm, atom by atom. This subworld is as varied and complex as any other level of being: a place unto itself. Its rules are neither those of galaxies nor those we see within the middle kingdom. This simple truth has puzzled many a would-be nanotechnologist.

It's hard to convey the strangeness of the nanocosm. What we know about our middle world is only a point of departure for understanding it. What Einstein said is true: The laws of physics are everywhere the same. At the same time, while basic laws don't change, their appearance can be wildly variant. Science and engineering get into trouble when they forget this. In fact, most of what they call "physical law" is not irreducibly basic. It's a jumble of empirical summaries, rules of thumb, and ad hoc adjustments that, just, well, *work*. Even quantum electrodynamics, whose predictions about the *sub*-nanocosm have so far proven accurate to 18 decimal places, uses a "normalization procedure" that cleans up bad data mathematically. When I took engineering, we called this type of thing Cook's Constant. High-energy physicists don't know why normalization works. Even its inventor could never explain it, which cost him no end of grief.

All this to say that in practical terms, the nanocosm is utterly different. To study it we must take a detour through another realm, that of measurement. Unlike the nanocosm it gauges, this world is synthetic and conceptual. It is a human product: intangible, yet as much an achievement as any building or book.

Le Système internationale des unités, SI for short, has its headquarters in a four-story building in Paris, France. The SI convention is responsible for maintaining and continually redefining the yardsticks by which science and technology record things.

The United States still uses the old, user-friendly, human-scale units of the British Imperial System: pounds and ounces, feet and miles. These units have become uniquely American since the Britain that evolved them abandoned them three decades ago. It's Britain's loss. Since the ancient units spring from everyday use they are easy to understand and apply, and so are very useful. One pound per square inch (1 psi) is a force you can feel; fourteen of them make up air pressure at sea level. A *pascal*, on the other hand—the corresponding SI unit of pressure—has the force of a hummingbird's sneeze. It takes 100,000 of them to make one atmosphere and 200,000 of them to inflate a car tire. The same reasons that make Imperial units natural, intuitive, and immediately useful also make them more poetic. As one disgruntled poet said, "Kilometers do not scan."

At base, though, even Imperial-U.S. units now define themselves by SI. Scientifically SI is all but universal; a pound is merely 454.00 grams. The whole structure of U.S. science uses SI entirely. A lab worker may buy her meat in ounces and drive to work in miles per hour, but she'll record her bench observations in meganewtons, kiloparsecs, and (yes) pascals. When push comes to shove, the whole might of American industry rests on a narrow street in Paris.

SI, as befits its origins in the Age of Reason, proceeds by what are called orders of magnitude. One OOM equals a factor of 10. A factor of one hundred is two OOM: $100 = 10 \times 10 = 10^2$. Three OOM is a thousand, and so on. SI considers every three orders of magnitude to be a step that's important enough to rate its own prefix. A *gram* is the weight of an eyedropper full of water, about a fifth of a teaspoon. A *kilogram* ($10^3 = 1,000$ g) is 2.2 pounds. A *megagram*, one million grams, weighs as much as a U.S./Imperial long ton. Up go the prefixes till you reach an *exagram*. At a trillion metric tons (1,000,000,000,000,000,000 g) an exagram approximates the combined weight of all buildings in the continental United States.

As well as reaching for the immense, SI delves in the opposite direction: down into the tiny. One thousandth of a gram is a *milligram;* one thousandth of that—a millionth of a gram—is a *microgram*. A thousandth of a micro-unit, one billionth of something, is expressed by the prefix *nano*—from *nanos,* classical Greek for *dwarf.* But whereas a human dwarf might be half the weight of an average adult, a nanometer is but one billionth of a meter. That's the diameter of a small molecule.

Here's an image, a la *Hitch Hiker's Guide*. If a nanometer were scaled up to the width of your little fingernail, then your fingernail would be the size of Delaware and your thumb would be the size of Florida. Yet the smallest manipulable element inside that monstrous hand, an atom of hydrogen, would still scale up to only one twenty-fifth of an inch. The nanocosm is a serious kind of small.

Small it may be; unknown it is not. *Higher, faster, better,* boast the Olympics. This could also be the motto of science, which constantly seeks to extend its understanding. But science adds other comparatives: *lower, slower, smaller, less obvious.* Johannes Kepler, the Renaissance astronomer who showed the earth revolves around the sun, put this endless quest for new knowledge in an arresting phrase. To be a scientist, he said, is to think God's thoughts after him.

In every new area of enquiry, science uses inference. There's not much *science* about science in a discipline's early stages, oddly enough. It goes mostly by guess and hunch. Scientists use their noses before they use their brains. A good scientist, said the Nobel laureate Gerhard Herzberg, can sense a pattern when only a few of the facts are in, and are badly distorted and swamped by noise, to boot. In this way, the first discoveries of the nanocosm came from logical inference based on indirect observation. At the outset of investigation, the tools for directly observing the nanocosm did not exist.

In 1808 John Dalton, an Englishman, concluded that every atom of a given chemical element was identical. These interchangeable atoms, he announced (to great skepticism from the scientific establishment), combine and recombine to create the infinite variety of compounds we observe at our everyday scale. Dalton's insight came nearly two centuries before individual atoms—or at least their outer electron structures—were directly inspected through a modern imaging device called the STM (scanning tunneling microscope).

Sixty years after Dalton, a Czech monk named Gregor Mendel noticed how certain traits in pea plants—flower color, for example—obeyed strict rules in how they transmitted themselves from parent to offspring. Since the long DNA molecule that instigates these miracles is only 2.3 nanometers wide, the best laboratories in the world were still 130 years away from imaging a living gene. But by an astonishing mental leap, Mendel had detected many of the overall structures and functions of heredity. Genes (he deduced) *must* exist with certain properties; otherwise his observations would have been different.

There were, and are, precedents for such imaginative star travel. Scholars knew that the world was round for two millennia before Magellan gave their theory a practical demonstration in the sixteenth century. Eratosthenes of Alexandria had deduced our planet's shape from the length of noontime shadows at different latitudes in Egypt. Similarly, scientists have known for generations that the nanocosm must exist. *Something* had to be down there like Atlas, holding the visible world on its back. Yet it's only in our own time, and especially in the past five years, that we've begun systematically to explore the nanocosm. Future ages will record the twenty-first century as the Renaissance of the Nanocosm, when the first great voyages of discovery were made into this bizarre interior realm.

While the word *nanotechnology* has gained wide currency, its use to mean something already in existence was initially premature. Even today the nanocosm has not generated much solid technology. It's about to; that's inevitable. But the bulk of it is a few years, and in some cases more than a decade, away. We've only begun to sail, chart, and record; we still haven't undertaken systematic trade or colonization. Today the nanocosm is like electricity in the age of Faraday, or heredity at the time of Mendel. We are still a long way from complete scientific explanations, let alone the robust economic sectors that these insights will generate.

Still, nanoscience has recently made such staggering gains that it is undeniably on the brink of a true nanotechnology. We have now mapped enough of the nanocosm to let us make educated guesses about the type of world it will soon support. These estimates range from the merely surprising to the wig-flippingly outrageous. Some very big changes in business and leisure are about to come to us by way of the very small.

Unfortunately, whatever does arrive will have to overcome a vast amount of hyped-up public expectation. Nanotechnology may be the first new technology that gained a large and vocal community of boosters before it was even close to existing. Twenty years ago while pursuing his doctorate at MIT, K. Eric Drexler wrote what some consider the first journal paper on advanced nanotechnology, envisioning what it might be. Dr. Drexler boldly foresaw a world of molecular manufacturing, where macroscale objects were assembled atom by atom by nanoassemblers the size of molecules. Ten years ago Dr. Drexler greatly expanded this initial vision in a full-length book, *Nanosystems: Molecular Machinery, Manufacturing, and Computation.*

Dr. Drexler's book is radical only in its subject matter—namely, making nanoscale machinery. Its engineering approach is classical, even cautious— bearings, couplings, even a section on how to make sub-microscopic nuts, bolts, and screws. Its vision is, in a word, macrocosmic. The book admits no difference in principle between building a half-mile suspension bridge and creating ten-nanometer bloodstream-cruising submarines with claw-like attachments to grip and place individual atoms. As an engineering environment, he assumes, the nanocosm is just another arena where engineers can apply known techniques.

On first reading, Dr. Drexler's 556-page book doesn't reveal itself for what it is: a highly detailed piece of speculation. Instead it seems so exhaustive, authoritative, and well-thought-out that it pre-answers all objections and equips nanotechnology with everything it needs to arrive. Certainly some books subsequent to Drexler assume this. The essay collection *Nanotechnology: Molecular Speculations on Global Abundance* appeared in 1996, four years after Drexler's tome. It, too, has an MIT connection, both in its contributors and in its publisher, MIT Press. (Though Boston has a conservative, stuffy image, parts of the place can seem like colonies of northern California.)

Nanotechnology starts out with a brilliant factual survey by its editor B. C. Crandall, but ultimately degenerates into bad science fiction. My favorite chapter is *Utility Fog.* In it, a moonlighting computer engineer (self-described as "married with two robots") rhapsodizes about a world whose air is literally packed solid with quadrillions of "foglets," micron-sized machines that answer voice requests and materialize whatever one wishes from thin air. Don't worry about breathing them in, he says— they'll squish aside and make room for lungfuls of air. *No problemo!*

In both this *faux* nonfiction and in outright science fiction, nanotech boosters proselytize their eccentric world like the pigs in *Animal Farm:* "Hearken to my joyful tidings / Of a golden future time." Nonfiction's gum-snapping zeal makes it more entertaining than most sci-fi.

DESPITE THESE imaginings by a fringe of boosters, there are strong signs that a workable nanotechnology is at last being born. Advances are occurring daily. While most of these are in basic nanoscience, business indicators such as total number of start-ups, IPOs, and pools of committed venture capital indicate that a viable commercial enterprise is emerging. Paradoxically, one of the strongest signs of this advance is the extent to which high-profile scientists, including Nobel laureates, feel the need to rebut the nanobooster brigade. It's significant that the doyens take such trouble to reject the boosters' claims. You don't see them put equivalent energy into debunking spoon-bending or UFO sightings. Nor do the senior scientists content themselves with simple rebuttals. Instead they produce soberly reasoned, thoroughly documented contrarian positions on how true nanoscience should proceed, and what legitimate nanotechnology it could engender. The fringe element's opinions don't matter and never did; but their frantic yapping has at last caught the attention of some reputable thinkers. Now we'll see some results.

It turns out that the learned opinions of the better scientific minds are more exciting than the blind faith of the whackos. This is not unusual. To the healthy mind excitement increases with nearness to reality, and reality increases with ironclad reasoning from a solid knowledge base.

Case in point: Richard Feynman titled his seminal 1959 lecture to the American Physical Society, the Speech That Started It All, "There's Plenty of Room at the Bottom." Science and engineering, Feynman said, should look inward to the nanocosm as well as outward to the macroscopic world. It was time we started making things not by carving away zillions of atoms in crude chunks, but by building up what we wanted with molecular precision, literally atom by atom. This is the point of departure for Drexler et al., and the rationale for their incredibly detailed imaginings of molecular machines.

Richard E. Smalley disagrees. Dr. Smalley received the 1996 Nobel Prize in Chemistry for his discovery of fullerenes. Consider the nanomanipulator, he says: a theoretical, molecule-sized machine designed to truck around individual atoms. This is an article of religion both in nanofiction and among the nanoboosters, and according to Dr. Smalley, it's a pipe dream. Based on what we're learning about the nanocosm, Dr. Smalley says, any device built to manipulate atoms on the nanoscale would necessarily have fingers that were not only too fat (i.e., too large and clumsy in relation to the atoms being manipulated) but too sticky. Carbon atoms, for example, can bond instantly to any matter that comes close. This electrostatic promiscuity makes carbon central to the products, processes, and molecules that we call life. But to Drexlerian engineering, carbon is a disaster. The instant a nanomanipulator arm touched carbon, it would become as immobilized as Brer Rabbit punching the Tar Baby. Other leading scientists have compared using a nanomanipulator with trying to assemble a wristwatch without instruments, while wearing thick gloves and with every part soaked in glue. There's not *that* much "room at the bottom," Dr. Smalley concludes. "Wishing that a waltz were a merengue—or that we could set down each atom in just the right place—doesn't make it so." Nano-sized manipulators are simply bound by too many ironclad constraints to work the way the boosters prefer to imagine.

Fortunately, that inviolable constraint doesn't matter to real nanotech. Even the elbow room we do have at the nanoscale is more than enough for several lifetimes of discoveries. Take, for example, the natural substances whose discovery won Dr. Smalley his Nobel—fullerenes, or (more formally) buckminsterfullerenes. They're named for the visionary U.S. engineer R. Buckminster Fuller, inventor of the geodesic sphere. Bucky Fuller's designs, Dr. Smalley showed, merely recapitulate what nature first did a few billion years ago. In nature, carbon atoms can spontaneously link themselves into geodesic spheres—the fullerenes or, more familiarly, "buckyballs." The buckyball carbon allotrope is abbreviated C_6O, for the 60 carbon atoms it comprises. It looks like a nanoscale soccer ball.

In his geodesic designs, Fuller unwittingly emulated nature, a process that nanotechnologists call biomimicry or biomimetics. Bucky's biomimicry, however, was wholly unintentional. In the macroworld, it just seemed like good design. It was—both artificially in the middle kingdom and naturally in the nanocosm.

There exist other, linear, nonspherical types of fullerene. Some natural carbon atoms spontaneously form tiny hollow cylinders with outside diameters of only one nanometer. Like many structures at the nanoscale, these carbon nanotubes (CNs) exhibit properties that seem bizarre or even contradictory to our middle-kingdom eyes. Aligned in a certain way, their atoms conduct electricity as effectively as copper. Aligned in a slightly different way, they are semiconductors. (Semiconductors are midway between electrical conductors and electrical insulators. This physical ambivalence makes microchips possible.)

The CN's surprising range of properties opens the door to computational devices measured neither in millimeters nor in microns, but in nanometers. At nanoscales, you don't need a cascade of countless electrons to make a counting device "flop" or change state, the fundamental operation in computing. A nanoscale device will flop when a single electron is fed into it. Or a single photon, if you want to run your nanocomputer not by hot, slow, old-fashioned electricity but by cool, fast, efficient light.

Soon a supercomputer's central processor could be smaller than a speck of dust. Most of even that tiny volume would be not brains, but "dumb stuff"—the leads and connections needed to import the nanoscale computation into our everyday world, and to carry down software into the nanocosm.

Dr. Stan Williams of Hewlett-Packard Laboratories' Feynman Lab thinks that "nanoscale computing will come on stream by 2006, just as current silicon-based technologies hit their theoretical limits of [low] size and [high] speed."

Carbon nanotubes have another striking property: structural efficiency, or strength per unit mass. They're like nanoscale wires. Their carbon atoms are so tightly bound to one another that CNs are extraordinarily resistant to being stretched or pulled apart.

Nature produces CNs only a few hundred nanometers long; scientists in the laboratory have already extended this by three orders of magnitude. Once the lengths of manufactured CNs reach a few hundred meters, we will be able to support structures the size of the Golden Gate Bridge by cables no thicker than a pencil. From such cables we will then be able to braid big guy ropes to stabilize a carbon-based mast a hundred miles high. This will ride out gales or hurricanes without undue swaying. If such an enormous mast were constructed as a hollow network and a simple elevator installed, we could launch satellites by walking them to the top of the mast and shooting them out sideways. The cost of launch would be cut by over an order of magnitude. This is speculation, granted. But it is founded on facts rather than on *a priori* reasoning, and it has a chance of coming true.

IN A SENSE, *all* technology is nanotechnology. That's because everything we use relies to some degree on the properties of matter at very small scales. A mirror is reflective because at the nanoscale, the metal coating behind its glass configures its electrons as an "electron gas" that returns incident photons better than a backstop returns tennis balls. And in some forms, nanotechnology has even been around for years. The soot put into vulcanized rubber tires to keep them flexible and strong across wide temperature ranges comprises particles of carbon only a few nanometers long. Cars have driven on these "butyl-latex-sulfur nanocomposite elastomers" for decades. The nano-soot is what makes tires black.

What sets apart today's nanotech from these early cases is intent. When we devise and fabricate today's nanocomposites, we know what we're doing. We're not just being empirical. Nanotechnology will soon let us bypass the substances that nature provides and start with a wish list of properties that a new material must have. We can then pick, choose, modify, and synthesize various molecules, creating stuff that meets our performance demands. High mid-temperature thermoplasticity in a dense, transparent solid? Strong magnetic properties in an inert, low-cost, sprayable coating? In five to seven years you'll be able to call a nanomaterial firm and tell them what you need. They'll make your substance to order.

Already nearing commercialization are new plastics that can hold a static charge long enough to attract and bond sprayed aerosol or powder coatings. Powder coatings have been applied to metal substrates (such as the galvanized sheet steel shells of stoves and refrigerators) for thirty years. Now, thanks to nanotech, it's plastics' turn. New nanofilm processes will permit efficient, low-cost coatings over plastics without the mess and waste of overspray. Anything that misses its target will be sucked back by electrostatic charge to bond in its proper place.

In fairness to the nanoboosters, their zany speculations have undoubtedly advanced the agenda of nanoscience. Life imitates art: By and large, we invent only what we first imagine. By piquing interest, first in the broader public and then in mainstream scientists, the boosters have advanced basic nanoscience and accelerated the commercialization of its discoveries. The boosters, bless their goofy hearts, have thrown open the doors to more disciplined imagination. In so doing, they have (however briefly) filled a real need.

While it will differ from our present world in various ways, some of them profound, the nano-based world of tomorrow will not be totally unrecognizable. For all its changes, life has one remarkable constant: human nature. This will prove an impassible roadblock to the scarier aspects of the boosters' imagined world. The barrier to much nanotechnology may lie outside the strictly technical, in what (to engineers, at least) is the mystical realm of marketing.

Take those blood-cruising nanosubs. I, for one, do not want any such nanotechnologically based, semi-intelligent "vivisystems" rooting through the natural gates and alleyways of my body, thank you. My blood vessels are my own. Multiply my case by four billion, and you realize that the nanoboosters' wilder predictions, even assuming against all probability that these prove feasible, will run smack against people's innate (and entirely understandable) fear of hosting new and unknown parasites.

What of the immediate future, say, through 2015? Even based on the sober, justifiable assessments emerging from mainstream science, the nanocosm promises to transform our lives by revealing new basic facts that we can turn into useful technology. This will occur not just in details, but in everything—how we work, play, live, communicate, and think. No, we won't breathe hosts of tiny nanomachines, now or ever. We won't achieve a world of telepathic, telekinetic, universal systems that make every wish come instantly true. Even as transforming a thing as nanoscience is still bound by physical and marketing limits. But singly and collectively, the nanocosm will transform us. It will not content itself with revolutionizing the grand things: economy and culture and democracy. It will alter, from the inside out, the myriad small details that affect us—how we stay healthy, how we spend leisure time, how we raise our children. The nanocosm that supports these widespread changes may not always be apparent, but perceived or not, it will be the agent of revolution.

CHAPTER 1

NANOWORLD 2015

*When sorrows come, they come not
single spies, but in battalions.*
— Shakespeare, *Hamlet*

GRIEF MAY COME as a cloudburst, but good news—especially that modern form of it called technology—has occurred to date as a steady, all-day drizzle. Futurists and historians, and their fictive counterparts the historical novelists and science-fiction writers, see this more clearly than most. As they consider where the present started, the long view is forced upon them.

Not much may seem to change in your day-to-day round of getting and spending, washing dishes, and going to school. But when you sit down and sum up all the tiny adjustments that an average modern decade brings, you realize just how far the leading edge of our planet's culture has gone—and how quickly it got there. Now the speed of such transformation is about to take another leap.

In August 1972, I dropped in on my great-aunt. This wonderful woman seemed the embodiment of changeless continuity. She still lived in the house her great-grandfather had built in 1832, and in which four generations of her family had grown up. Many of her relatives, from newborns to nonagenarians, slept under grass in the neighboring churchyard. She never married, instead tending old family—grandparents, parents, finally sisters—until she herself grew old. All this occurred beneath the same slate roof. The oak planted when the house was built was now a patriarch nine-feet thick at its base.

As my aunt and I stood in the twilight outside her front door, I saw a strange expression cross her face. "I was just thinking how the world has changed," she said. "There were no motorcars when I was born. That garage was first built as a horse stable. This house had no electricity for its first sixty years, and no telephone for its first seventy. Now I've lived to see people on the moon. You young people take so much for granted. You have no idea how it was back then." How was it? I asked. She took her time answering. "Different," she said. "It was utterly, utterly different."

She was dead six hours later: That night she slipped away in her sleep. Maybe that's why her words stuck in my mind. They had, and have, the force of prophecy: a wise old woman's final words. But as I began my research for this book, I realized she was on to something with wider implications: something subtle and profound. She had put her finger on the scope of change.

It goes like this. Political revolutions happen suddenly. On Monday a despot sits entrenched, on Tuesday there's a local food riot, on Wednesday the disturbance spreads—and the old man's dead or in exile by the weekend. The causes may be ancient; the effects take place in a flash.

So far at least, revolutions in science and technology have taken longer to occur. Og the Troglodyte didn't wake one morning and say, "Hey! Let's domesticate some animals." The word *revolution* was applied to the Neolithic Age only after the fact. The Industrial Revolution took a century and a half; the Agricultural Revolution, two or three millennia. The extent of the change, its economic and cultural impact, was hardly apparent at the time. In the words of Andrew Marvell, it was "vaster than empires, and more slow."

This is true even today, when the pace of technical change has increased. Telephones, cars, the cotton clothes and sanitary sewers that ended the cholera and typhus plagues—the triumph of these things was steady but slow. Even computers were not enthroned overnight, borne to the palace on the shoulders of ecstatic mobs. They achieved supremacy, if you'll forgive the pun, bit by bit. Only in retrospect was a technical revolution apparent.

Yet revolution there was; and to see its extent, you have only to think back. Anyone my age (I was born in 1946) can, with a little reflection, attest by direct witness to how far we've come in the last fifty years. *Return with us now to those thrilling days of yesteryear!*—as the radio announcer intoned at the start of radio's Lone Ranger Hour. I have memories from 1952 of my father installing a kitchen floor with rolls of that synthetic wonder-material, linoleum. 1952 was a year for miracles. The same kitchen reno added a pot light, a device that recessed a 60-watt bulb into the ceiling so that no dust-catching fixture hung down. Even more magic was that year's mass conversion of every electric appliance in Ontario from 25-cycle alternating electric current to 60-cycle AC. We didn't have the term *Hertz* then; it wasn't in use. We said *cycle*, for cycles per second. For three million homes and businesses, lamps no longer caused eyestrain with their weak, dim flickering. Instead they shone with a pure, clear, steady light that made reading enjoyable even after sunset.

The year of 60-cycle power was also the year of the rotary phone dial; the party line; the school desk with a built-in inkwell, into which I dipped a steel-nibbed straight pen without a reservoir. It was a year of cars without seatbelts or pollution control; of "rapid cal" drills by which we schoolkids practiced multiplication, addition, and division in our heads; of human runners who hand-delivered unpunctuated telegrams: URGENT NEW YORK SOONEST STOP. There were few or no handheld calculators, commercial jets, or radial tires. The handful of computers that existed were monstrous vacuum-tube mainframes that took up rooms. Few women worked except as secretaries, clerks, and teachers. Those who did work were either young and unmarried, or else very old.

Next door to my grandfather's big uninsulated home, a middle-aged neighbor died a lingering and painful death from kidney failure, renal dialysis and organ transplants not having been developed. Granddad had health worries of his own. He passed small kidney stones and was surgically slit from groin to gullet for the big ones; the stone-pulverizing lithotriptor was thirty-five years away. Worse for him and his family were Granddad's undiagnosed transient ischemic events. The term hadn't been coined yet, but the effect was real enough. A series of mini-strokes bent my grandfather's formerly extroverted personality until he became as bitter and paranoid as a biblical patriarch. (On reflection, that's probably what happened to Noah and his ilk, too.) Stroke was epidemic among middle-class, middle-aged men like my grandfather. Even if your heart held out, your veins and arteries wouldn't. After a life of booze, stogies, edible grease, and little exercise, there was no escape.

Society had as many undiagnosed problems as medicine. Cops winked at wife-beaters and let drunks drive if they promised to drink coffee.

Teachers routinely assaulted their students with wooden paddles and rawhide straps. Playgrounds were lawless jungles of bullying and harassment. It was either legal and official, or else condoned.

URBAN NORTH AMERICA, 1947—7:30 A.M. Joe Johnson's wind-up Westclox, which has clanked like a metronome all night, lets loose with a twin-bell Klaxon that would wake a corpse. Joe slaps a push-knob to kill it, scratches his belly through flannel pyjamas, and shuffles downstairs. He puts a battered tin pot of coffee on the stove to perk, hits the shower, takes a pair of hairbrushes to the half-inch fuzz atop his head, and blade-shaves into a steamy mirror, sticking bits of toilet paper on the cuts. Joe's wife is up, too—only the boys sleep through that alarm—and is cooking breakfast. Joe dons his brown serge suit and red tie, sips his coffee, and waits for his wife to serve his eggs and bacon. After breakfast he shrugs into his topcoat, jams on his second-best fedora, and gives Joanie a quick peck on the cheek. Then it's out to the car and two miles to the downtown office where he manages Payables. Today Joe has a sales rep coming in to show him a new filing system based on color-coded cardboard. He's set the meeting for 11:30 so the guy can buy him two or three beers at lunch.

CUT TO 2003.

Joe Johnson II wakes at 5 A.M. when his cellphone chimes. It's New York. Cincinnati calls early because they forget the time difference; New York doesn't remember and doesn't care. The place that never sleeps makes sure the rest of the planet doesn't sleep, either. Still, Joe is polite; a deal depends on it. He talks as he walks naked into the kitchen—*Uh-huh, uh-huh, sure*—and pours coffee from a machine. His live-in girlfriend is in Singapore and doesn't know when she'll be back. Joe's glad of that. He'll need the extra time to complete his HTML demo before catching Friday's plane to Kennedy.

Joe powers up his home office. E-mail full as usual. The spamsters have found another route around his filters. Click, drag, group delete. There's something from his father. Get together some time, things to tell you, yada yada. Sorry, Dad, can't make the burbs for at least a month. Joe checks his watch. If he hurries he has time to hit his club and still make the office before seven. He's glad he spent the extra hundred K to get a one-bedroom condo downtown.

In the future when you recall 2003, you'll have the same shock of unfamiliar familiarity that you got reading about 1947. This year will seem like the Middle Ages. You'd feel that retrospective alienation and distance even if change—both technical innovation and the social adjustments it catalyzes—were as slow as it was over the last half-century. But the rate of technical change is increasing; a greater contrast is in the works. A.D. 2003 will seem antediluvian not in fifty years but in fifteen. For the first time in history, a technical revolution will approach the abruptness of a political event. This is the promise of nanotechnology, and will soon be its legacy. No one in any age has heard, seen, or felt anything like it. But you will.

CUT TO 2015.

Joe Johnson III turns forty today. He's not aware of it because he doesn't know what day it is, or even if it is daytime. It isn't daytime; it's 4:00 A.M. and Joe hasn't slept for thirty hours. But then, he doesn't have to be aware of time, any more than he has to yield to his fatigue. Neither situation bothers him. A cocktail of time-release medication in his bloodstream keeps him as alert and refreshed as if he'd just snored through the last ten hours. Twice a day he lies down for fifteen minutes so his R-brain can reorganize and file data in its ancient way, by dreaming. Then he's up and at it again. The medication isn't the old hit-and-miss variety, either—chemicals dumped into the body in the pious hope that some of them will end up where they should. This stuff always works. It's ferried by synthetic molecules called dendrimers, each 25 nanometers across. The dendrimers release their tiny cargoes of fatigue suppressant only when they're safely inside a target cell in Joe's midbrain. Dosages have dropped from milligrams to nanograms in the last five years alone.

Right now Joe is using his nano-medicated alertness to watch paint dry. Actually it's dry already, even though he rolled it on only last night. Every square inch of his bedroom, even the floor and ceiling, is covered with the stuff. It's a colloidal dispersion of carbon nanotubes that react to polarized current by changing pattern and color. Joe doesn't need a TV screen to work. His whole room erupts with data. The paint he applied is a roll-on display screen that's a hundred microns thick.

"Just a minute," he says to his partner, whose giant image grins down at Joe. The partner is a young man wearing sunglasses. He also sports the most fashionable new skin shade, ice-melon green. The guy is so cool that Joe suspects *he* is actually an *it:* an avatar, projected for human interface by a powerful AI, or artificial intelligence, program. Real or synthetic, the guy knows his stuff. He/It and Joe have made themselves good money for the last fifteen months. That's an epoch in this age of instant partnerships.

Joe has no idea where the guy, or thing, is based. Now and then there's a brief response delay, hardly more than an eye blink. This could be a live guy pondering. Joe, like all humans, pauses from time to time to think things through. Alternatively, the pause could be the transmission lapse of a photonic wavefront. It takes measurable time even for light to cross thousands of miles of TIR fiber or else reach a geosynchronous communications satellite 40,000 klicks above the earth. Joe knows his partner's URL but not his physical location, assuming that phrase has meaning anymore. The man, or the machine, may be in Antarctica for all Joe cares. What matters is, they've set up a machine farm to make consumer goods.

"Show me the sales forecast," Joe says. He brushes the stubble on his nose, feeling a flicker of annoyance. The pilatory he took last week was full of simple ten-nanometer machines that sought out dormant hair follicles, attracted capillaries to give themselves a blood supply, and then proceeded to extrude an endless keratin cylinder—a hair, in other words. That gave Joe a nice thick head of hair. Great, but it was too powerful. Some of the follicles it kick-started were in Joe's nose, which merrily began to sprout coarse black hair. Now Joe looks like a werewolf even after he's shaved. He has to adjust his outgoing signal to clean up his image. He wonders: Do AI avatars have a sense of humor? Probably no more than any business type. Years ago someone defined an MBA as a computer without the personality.

Joe swivels to examine the wall behind him as charts spring into life all over it. "Good saturation," he says. He means market presence, although the richness of the graphic colors is good, too.

"We dominate the market," his partner says. "We'll gross six million this week." Joe smiles. Even here in San Francisco that's ten years' rent. Still, his partner thinks they should go on to develop other things. "You know what product life-cycles are like. We've had DustGone on the market for nine weeks and it's getting stale. We need a winner to replace it."

"Well, that's what this meeting's for." Joe yawns, glancing leftward at the time display. Two hours till his next REM break. DustGone, which they brainstormed back in April, took off the week it went on sale. Each sale package contained two million 70-nm robots in an escape-proof bag, programmed to do three things. First, on being dumped into a household, the nanobots self-replicate to a predetermined limit—a hundred trillion per thousand square feet of living space. Second, they then prowl the domicile, killing every microparasite they come across by tunneling through its skin. The nanobots are powered by an ATPase-B flagellate motor, like the one that moves *E. coli* around, but modified so it works without water. Third, the DustGone nanobots, true to their name, break down dust into carbon dioxide and trace elements, releasing them harmlessly into the air. The product's ads call it *a live-in maid in a plastic package*. To a society rarely at home and too busy to clean, it's a godsend.

The only problem is, the nanobots are too efficient at self-replication. A purchaser can wait a week, then sell a bagful of enriched dust to any number of his friends at a thousand bucks a pop, one-tenth the price of an over-the-counter package. DustGone is burning through its life-cycle in half the time Joe and his partner intended. It's becoming a commodity, tradable (and traded) in the black market. They have to find something else to fill the gap, fast.

Luckily, nanotech means there's no shortage of possibilities. The same dirt-cheap Chechen machine farm that breeds DustGone nanobots can easily be reprogrammed to make other things.

Joe III swivels back to his partner's green-skinned image. "Let's recap what we know about the basics. The world currency is U.S. dollars. That means the whole globe's functioning as a unified economy, whatever national barriers are officially in place. And that world economy is driven by—what?"

"You tell me."

"Two things. The first is transportation; people have to get around."

The partner frowns. "You and I don't commute anywhere."

"Not for work. But what's the other thing that drives the world? Come on, you gotta know this."

The partner shrugs. "Goofing off, I suppose. Okay, leisure."

"Correct, leisure. Consumers rule. They're the only money source worth bothering about. Industry doesn't matter unless it's hooked up, directly or indirectly, to consumers. And what do consumers want?" The partner shrugs. Joe continues: "To get to where they can goof off, right? The golf course, the beach."

"So what?"

"So there's still a high demand for stand-alone powerplants. Most of them are for personal transportation. Look at the stats."

The partner looks over Joe's shoulder, presumably at a data readout of his own that Joe can't see. If this is AI, it's pretty sophisticated. Then he/it nods.

"You're right. Mostly vehicles, some power generation. What's your idea? Where do we fit in?"

"What about having the farm grow diesel engines?"

There's that brief hesitation again. Either the partner's ruminating, or the AI that fronts him as a human-machine interface is accessing its files. The partner says: "Old internal-combustion, right? No spark plug? Piston squashes an air-fuel mixture till it ignites?"

Joe nods. "You got it."

"It's too dirty. Jeez, you couldn't make that in Afghanistan. Or you could make it, but you couldn't import it anywhere that could afford to pay for it. Nitrogen oxides, particulate matter, carcinogens, the damn things have been banned for years."

"Not anymore," Joe says. "I just found out US-EPA is changing its regs. You can clean up a diesel so it blows nothing but water and carbon dioxide."

"Just like a fuel cell?"

"Just like a fuel cell. Couple of profs at Indiana State figured out how. All it takes is a nanocatalyst."

"A what?"

"A material we can get designed. It acts as a catalytic surface. All that ugly diesel stuff breaks down in a millisecond. You could breathe the exhaust and never cough." *Assuming you breathe,* Joe thinks.

The partner chews on that. "There's a big fuel-cell lobby not gonna like that," he says at last. "They must have invested fifty bil in proton-exchange membranes the last thirty years."

"So what? They missed the boat. You know how Mark Twain went bankrupt?"

Pause. Search. Then: "He invested in an alternative to the Linotype machine."

"And which technology won out?"

"Linotype. By 1920 there was no other practical way to set hot-metal type."

"Moral?"

The partner grins. "General Motors're gonna hate us, Joe."

"They're not the only ones. A catalyzed diesel could burn sludge, you know? Used oil, stuff so crude it hardly needs a refinery. You could use it the way it bubbles out of the ground. That drops demand for oilco product right there."

The partner raises his/its eyebrows, types a note. "Sell oilco stocks... You sure this thing is workable?"

"I've read the research papers. Internal combustion, sparkless but without noxious emissions. I know, I know, it's counterintuitive. But we could ramp up fast. Next month we could be growing a thousand engines a day that harness the effect."

"Grow 'em from what?"

"Material we'll design. Have designed, rather."

Another note. "Specs?"

"Must withstand very high internal pressures and temperatures. Good R-factor—must be a great insulator. We'll need a cylinder wall two millimeters thick with 1,400 Celsius and fifty atmospheres on the inside and room temp and fifteen PSI on the other. Plus very low stiction coefficients."

"Say what?"

"Stiction. Surface values for standing and kinetic friction values at the nanoscale."

"Well, if that's all we—"

"Uh-uh, there's more. Mechanical characteristics—compressive and tensile strength in the order of 200 kips—zero ferromagnetic properties." *Kip* means thousands of pounds per square inch. Joe's partner must know that because he/it doesn't question it.

"Sounds like a ceramic," the partner says.

"Substance doesn't have a name, it's never existed before. Noncrystalline, nonceramic, nonmetallic . . . .What's our company name?"

"JOE-X," says the partner.

"Call the stuff Joxite. We'll have it custom-designed. Piece of cake."

"How can we farm this stuff? What procedures?"

"We'll check out a bunch of ways and take the best. We could start with a solid block and have the 'bots bore and mill it till we get our final shape. Or we could injection-mold it. Or build it up *in situ*—grow it in a tank. I have a line on a private lab that can get us the optimum method."

"I thought nanoassemblers couldn't exist," the partner says. "Thought they were science fiction."

"They are," Joe says. "That's not what we're after. We want dumbots. Chemically, dumbots do one or two things only. Think of 'em as synthetic catalysts."

"Diesel—engine—farm," the partner says as he makes a note. If this is an avatar, Joe thinks, it's a good one. "Right! I like it. What else you got?"

"Involves leisure again. We will now have lots of people driving cheap, fuel-efficient engines, tearing off to enjoy themselves. Where they gonna go?"

Shrug. "Trout streams?"

"Golf courses, mostly. No trout left. Golf's growing, has been for thirty years. Look at the demographics."

"So people play golf," the avatar says. "Whadda we supply?"

"Better equipment," Joe says. "They're playing an old game, but they want newer gizmos to play it with."

"Be specific."

"Sure. Clubs that hit well now are heavy, they use mass to impart momentum. That makes 'em hard to lift."

"Doesn't matter," the partner-image says. "There isn't a course in the world that lets you hump your clubs. They all insist on power carts. Maximizes user throughput. Though we could probably sell our clean diesels to the cart people, too—"

"I don't mean *carrying* the clubs," Joe says. "I mean *swinging* 'em. You want to hit long now, you need a heavy club. What if we went the other way and made a club that weighs next to nothing? So easy to swing that you could generate oodles of specific impetus if you put your ass into it?"

"Featherweight?"

"Way less. Zero weight, practically."

"How would we do that?"

"Make woven clubs," Joe says.

"Woven? What kinda thread?"

"Buckytubes," Joe says. "Carbon allotrope, one of the element's basic forms. Basic buckytube is single-walled with a skin one carbon atom thick. One nem in diameter. We could weave double-walls, skein those

into ropes, skein the ropes into cables. Weave the cables into a club."
Nem is the new slang for nanometer.

"How big are these cables, Joe?"

"Hundred nems or so. Tenth of a micron. You'd end up with something that feels and looks like smoke. Eighty, ninety percent transparent, you can see the ball through them."

"Weighing how much?"

Joe shrugs. "Hundredth of an ounce maybe. Four grams. Mass of a headache pill."

"Might not work if it's windy."

"But it might. We'll test some prototypes before we mass-produce."

"Why weave a whole shape? Why not get by with singlestrand? Lots easier to make, Joe."

"Because it's too dangerous. One strand would hit you like a surgical scalpel."

"A what?"

"Scalpel. Super-sharp knife doctors used to use to chop into people. Remove things, stitch in organs."

"You're kidding me."

"No, they used to do that once. Now they go through a sweat pore; the skin's full of holes a hundred microns wide. To nanotech that's a city gate, it's something you could drive a car through. But they used to cut and sew people like cloth." Joe smiles as his partner flinches. "Anyway! You do the shaft of a golf club as a single buckytube, and it becomes a weapon. Carbon-carbon bonds are the strongest they know. It would take a couple of freight engines pulling opposite directions to rupture one. A thread one atom wide, too small to see—we could make that. But it could cut off your leg before you felt it."

The partner grins. "Gives a new meaning to golf slice, doesn't it?"

"It does. So I say, beef it up. You won't add much weight. Whole club'll move like a breath of air. It won't even make a sound when you swing it, 'cause the threads'll cut through the air like so many knives."

"Not even a *whoosh*?"

"Not even that. But make your club like an armature, and it's usable. Like those old handles on cast-iron fry pans, you know? Spiral shape, feels solid but it insulates you from the heat. Very, very low weight."

"Any other ideas?"

"That's it for me."

"Okay, I got some thoughts," says the partner. "How about snow-boarding? That's the big thing for the under-thirties. Can we give 'em something there?"

Joe says, "Let me think." He wonders if his partner worries *he's* a machine; it makes him grin. "Okay. Current boards are clumsy, one or two centimeters thick. Heavy to lug around, unresponsive to jink when you use 'em. What about we engineer something as thin as paper? Travel with it folded. Or better yet, in a roll. Unroll it, rigidize it, use it, pack it up again when you go home. Also ultra-low-friction—you'd need to prac-tice to stand up on the flat."

"That possible?" The partner raises his/its eyebrows.

"They're already using it for car bodies. We mold in a capacitance device that tunes stiffness by modifying intermolecular bonds . . . snow-board Viagra, dial your rigidity. Only one problem. Remember what I said about the golf clubs? Can't make 'em too thin or they act like knife blades. Make a board five hundred nanometers thick, and you do *not* want to get hit with the thing. Even traveling at ten nems an hour, it'd go right through you. You'd have ski slopes full of razor blades."

"What's a razor blade?" the partner asks, and Joe tells him. The part-ner asks: "Can you safeguard that?"

"We could have a warning system. A screamer that you hear only if you're in the flight path. You hear nothing, you're okay."

"No news is good news, huh? Leave that with me, I'll check out the legalities," the partner says. "One thing about a directional alarm—it would avoid noise pollution. Whole snow hill would be screaming otherwise."

"Can you run up our accounts?" Joe leans forward to examine the pat-terns that spring up. "Not too shabby. Where'd you get those interest rates?"

"Little S&L in Venezuela. Came across it last week."

"I'm glad we launched our own satellite," Joe says. "Best fifty mil we ever spent."

"We'd be crazy not to," the partner says. "Gives us all the data we need. Plus it's dead simple since they put up the Tower."

Joe nods. The Tower is a hundred-mile buckytube mast, anchored in Ecuador and extending above the atmosphere into space. A mechanical crawler traveling at walking speed moves up the mast and takes a satel-lite to the top within a day. A short blast from a solid-fuel rocket booster and your satellite's in a low earth orbit stable enough to last a century.

Capital outlay's a bummer, but they make it back twice a month in data-acquisition fees.

Joe stands and stretches; time for his treatment. "Anything else?" he asks through a yawn. His partner shakes his/its head. "Call me tonight. Six my time," Joe says. The partner nods and disappears; the room walls reassert themselves, glowing like sunrise.

Joe goes to his kitchen, pops a pill, and takes a shower. In ten minutes the gold nanospheres in the capsule pass through his bloodstream and home in on the metastasized melanoma in his lungs. Joe lies beneath a heat lamp while a timer counts five minutes. By lunch tomorrow his cancer will be gone, broiled by infrared radiation that the nanoparticles have focused. A real nuisance, cancer. Every time Joe gets the thing it costs him twenty minutes. And more than ever, time is money.

CHAPTER 2

NANOSCIENCE
TRENDS in WORLD
RESEARCH

VIFFING

SAY YOU'RE A galactic intelligence from beyond earth: a physical scientist. You come across earth in one of your periodic expeditions, and settle in to study it. The big patterns are immediately apparent. Earth is a planet, revolving at a nearly constant distance from a G_0 yellow dwarf star. It's almost perfectly round and rotates slowly about an interior axis. Its temperature varies, both with place and time—it gets colder the higher above the surface you go, or the closer to the poles. Anywhere outside a narrow band about the equator, there are seasons. The axis the planet spins around is raised 67 degrees above the surface of the imaginary plate described by its solar revolution. You deduce, correctly, that this axial tilt lets sunlight strike earth's curved surface more or less obliquely. A season is warmer whenever earth's axis points toward the sun, colder whenever the axis points away. In the buffer seasons that separate Hottime and Coldtime, temperatures are lukewarm.

Then there's the weather. You're conducting your investigation from earth's single moon, a ridiculously oversize mini-planet whose diameter is nearly a third that of earth. The moon is typical of the celestial bodies you've seen over the centuries. It's dead. Nothing's happened to it since its formation several billion years ago. But earth, now, earth is ridiculously different. It's as active as a living organism, and it's totally surrounded by an envelope of transparent gas. Powered by sunlight and that planetary rotation, drawing on heat stored in earth's vast oceans and influenced by crinkled landforms, this atmosphere swirls ceaselessly. Vast chunks of air parade across the face of the globe, ferrying with them clear high-pressure air or watery, low-pressure storms. High-voltage electrical discharges sizzle among cloud forms, or zap from clouds to earth. When your instruments show you these great sparks are hotter than the surface of earth's sun, you *viff* in amazement, waving your flagella wildly. Truly an astonishing place.

But there are wilder surprises in store. With the broad parameters noted, you turn to your more sensitive remote probes. Spectrographs show you which compounds and elements dominate the visible parts of earth. That restless atmosphere is mostly biatomic molecular nitrogen, N_2. There are traces of elements that keep to themselves and don't tend to combine with anything—argon, krypton, neon. But what's this! *Oxygen?* Oxygen is deadly poison. It's a brutal electron thief, ripping away the outer shell of almost every element it runs into. Earth's atmosphere is *twenty percent* oxygen. You tap your instruments, too surprised even to *viff,* but the readouts confirm the assessment. Earth swims inside a planetary bag of poison gas.

That's the beginning of a parade of miracles. Oxygen isn't confined to the atmosphere. It's everywhere. Combined with carbon and calcium, it constitutes the limestone atop the planet's coldest peaks. Bonded to silicon or potassium, it makes up earth's hottest deserts. It's dissolved in rivers, lakes, and oceans as O_2. Earth's initial, striking beauty is misleading. The whole planet is a corrosive hell.

You realize that to understand this crazy place, you're going to have to shrink the scale of your examination. You've been looking at gross features. Now it's time to come down an order of magnitude in scale. Earth measures twelve million meters through the middle. (Xenophysicists use the metric system because it was really invented in the Andromeda Galaxy, a fact that will not surprise Americans.) Earth's surface land masses can be fourteen million meters across—big enough to wrap halfway round the planet. Interesting, sure. But none of it explains where all that oxygen fits in.

Down goes your inspection level by another factor of ten. Big weather systems like hurricanes show up, a million meters across. But while they move oxygen around by the gigaton, they're not the source of O_2 production. They're big, dumb cargo transports. Same at the next order of magnitude: 100,000 meters, the scale of cordilleras and archipelagos. They're cool too, but they don't explain the oxygen.

At the 10,000-meter scale comes a knockout find. New features suddenly emerge, utterly unlike the crystalline regularities of natural structures. Quantum magnetometers show intense electromagnetic fields around the new features. You decide you need an even closer look.

At the 1,000-meter scale, the sprawling features resolve irregular grids. You crank your scale to one hundred meters, nearly the limit of your sensors, and stub your *snurt* on even more surprises. Here and there are hot sources of thermal neutrons and neutrinos, which coincide with the most intense EM fields. Is that fission power production? Are these artificial structures?

Drop to ten meters, one-millionth the scale of your first examination. Ships appear; trains; aircraft; waves of surface traffic. Artificial vehicles! A genuine civilization! By now, you're so excited you're vibrating. Everyone in basic research knows the feeling, a high that combines the jolts of public adulation, games, sports, victory, drugs, gambling, sex with someone gorgeous, and a big tax refund. Intellectually or emotionally, nothing compares.

The final power of ten: one meter, a ten-millionth the scale of your initial look. Bingo! Fixed organic structures with green foliage that take in carbon dioxide and emit molecular oxygen. And here, there, and everywhere: mobile entities that direct the traffic, and make the vehicles, and plant and cut the big sessile things. The mobile critters take *in* O_2 and give *out* CO_2. The fixed and mobile forms use one another's output in an endless feedback loop. And more important scientifically, an observation first made at the planetary scale finds its explanation seven orders of magnitude below the global. Damned good thing, too—the instruments are on their final stop.

As you pack up and return home, you cap your splendid haul of data with three baffling questions—*What is oxygen, anyway? Why doesn't it kill these creatures? How do they harness it?*—and a single recommendation. *Fabulous data at the millimeter scale. Need instruments with finer resolution!*

In setting out this discussion, I've gleefully disregarded novelist Robert Sheckley's excellent advice: Don't pick the analogy, it'll bleed. Sheckley is right. It's almost always unwise to argue from analogy at length. Besides, it's just a literary trick to compare nanotechnology with planetary science.

Or is it? I'm a contrarian. I believe our current knowledge of the nanocosm, and our current approach to it, is precisely like an imaginary alien studying earth. It goes beyond analogy, simile, approximation, or metaphor: It's an exact parallel. We stand outside the nanoworld and peer down to examine it, just as our imaginary alien friend peers down at earth. More to the point, we are just as likely as she/he/it to find answers by steadily reducing the scale of our investigations. This conclusion rests on a startling fact. It's true, as Einstein noted, that "the laws of physics are everywhere the same to all observers." That's the most compressed expression of General Relativity that's possible without mathematics. And yet—*pace* Uncle Albert—the *manifestations* of those laws that an observer sees are anything but constant. They vary wildly with the size of the observer. No, laws don't change. But how they reveal themselves sure does.

MATERIAL WORLD

Like all frontiers, a scientific frontier is a border zone: a dim, mysterious landscape where one thing becomes another. If you're a scientist, you want, nay lust, to examine this conceptual interface with ever greater depth and precision. Interestingly enough, nanoscience has recently identified a material interface that exactly corresponds to its ideological one. This material frontier is proving to be the biggest single means to advance our knowledge of the nanocosm.

I say *material* interface advisedly, for it is on the stage of materials science that most other nanosciences are converging. MatSci is a long-neglected discipline. It's the reinsurance sector of the scientific world: necessary, steady, lucrative, and utterly dull. Since it began in mining and metallurgical studies a hundred years ago, it has consistently provided fast, substantial economic returns that have repaid its costs many times over. Thanks to materials science, jets stay up and buildings don't fall down; trains roll cost-effectively and car engines last for years. But while well-funded and solidly successful, it's kept a low profile. Faced with a brilliant high-school student, few guidance officers recommend MatSci. Instead, they suggest the sexier disciplines: biotech, high-energy physics, drug discovery. Materials science has traded glamour for respect.

Yet in the last two years the ugly duckling has grown decidedly swan-like. Not only does a vast amount of nanoscience come to a sharp focus in MatSci; many new commercial applications rest on a knowledge of nanoscale materials. MatSci's finally been asked to dance.

MatSci's new face finds a human parallel in Dr. Doug Perovic, chair of Materials Science and Engineering at the University of Toronto. Brilliant and accomplished, starring in a discipline that is itself a rising star; young, lean, eloquent, good-humored, and unburdened with ego, Doug Perovic is out of most folks' league. You meet people like this now and then, men or women with everything. They have so much we lesser mortals aren't even jealous. According to a Japanese *tanka*, hills envy hills that are slightly higher. A molehill knows it can't be Mt. Fuji. It doesn't even rankle that Perovic looks like Myron's *Discus Thrower*, without the discus.

Perovic *Fuji-san* shows me into an office full of bright spring sunshine. The room dates from forty years ago, when academics—especially chairs of departments coordinating hot new fields for big, first-rank schools—had major cubic footage. This place is the size of some restaurants: five hundred square feet of floor with a ten-foot ceiling. Despite its size it still seems crowded. Perovic has packed it with books and papers, and computer equipment is stacked two feet deep on every horizontal surface. Perovic sweeps a space clear on a big meeting table and drops his raw-boned frame into a standard-issue chair that suddenly seems undersize.

"Nanotechnology?" he says, in answer to my question. "There's not much real stuff yet. It's coming, but it's not here. What we have is basic science, nanoscience. Before we get the technology we've got to understand the fundamental processes at this small a scale. That's where materials science comes in." Can he give a definition of nanoscience? "Anything concerned with features below a tenth of a micron—100 nanometers. Microtechnology in both materials and electronics already overlaps that threshold. Some microchip elements now range around eighty nanometers." Hasn't the nanoscale always existed? "Science has assumed and even described nanoscale features for over a hundred years, true. But now we can observe such things directly. More than that, we can manipulate them. The scanning tunneling microscope (STM) has generated and regenerated more careers than anything else in recent memory. It's burst open the doors to the—what did you call it?" Nanocosm. "Yah, I see. Microcosm, nanocosm . . . Okay, so this thing, the STM, blows in out of nowhere. It completely rejuvenates my own area, surface microscopy. Now we don't have to guess, or assume, or extrapolate; we *know*. We see things down to one nanometer resolution. Sometimes to one angstrom, which is ten times better. That's the size of atomic hydrogen, the smallest atom there is."

As well as nanoscale features, Perovic tells me, true nanoscience must by definition (his definition) concern itself with properties that are noticeable or even dominant at the nanoscale. "In one sense we've had nanotech for decades. A car tire is black because it contains trillions of nanoscale carbon particles. Carbon-sulfur bonds created by the vulcanizing process keep tire rubber flexible over a wide temperature range." Still, Perovic says, this older nanotechnology was light-years away from what's going on today. "The past stuff was strictly empirical. It worked, but nobody knew how. Today's nanoscience can see what's going on, then try to come up with deep explanations and engineered improvements."

One result of this change is the coming together of a vast range of previously walled-off disciplines. "Biology is crucial to nanoscience. So is physics, so is chemistry. So are the engineering disciplines, every one of 'em. The nanotechnology that's emerging from the basic discoveries requires several fields working together. I think the whole notion of distinct scientific fields is breaking down. In a hundred years people will shake their heads at how we old guys treated molecules and atoms differently whenever we changed viewpoints. When we stressed electron interactions, that was 'chemistry.' When we stressed self-assembling structures, that was 'biology.' Now we're starting to see how all those various approaches are really a single thing. We've just been looking at it in different ways. Seen in that light, a scientific or technical discipline is nothing more than an artificial construct, a tool. It's arbitrary, almost.

"All science studies matter and energy, right? It doesn't matter if you call yourself a physicist, a chemist, or a biologist, that's what you do. A whole range of disciplines is collapsing into a new structure, a nanostructure. Because at the nanoscale, matter looks and acts the same. Nature is one, a unity. And so is nanoscience."

That being said, Perovic singles out one interface that's generating a particular amount of excitement: his own. "To me," he says, "most nanoscience is materials science." And within MatSci there's a subsector that's the most vital of all. Perovic calls it *bio-nano*—the study of living systems at the nanoscale.

I put down my pen and look at him. Something's been bothering me, I say; I'd like your views. For over a generation, the tenor of discussion in science and technology has bordered on the hubristic. Disease will be eradicated. Cancer will be conquered. Near-space and the solar system will be colonized. Now, according to Eric Drexler and his ilk, nanotechnology will make us omnipotent. By performing mechanical engineering at the molecular scale, we will dispense with messy biology altogether. Nothing will be beyond us. We will become as gods, mastering time, space, causality, and even death. Is this view of things believable?

Perovic rolls his eyes. "There's so much *shit* out there, getting in the way of real nanoscience. No, we're not going to become godlike anytime soon. Every new discovery makes any scientist worth the name humbler, not prouder. Molecular manufacturing? Self-assembly? Designer materials? Nature, my friend, has been doing all that for billions of years. My God, there's so much we don't know. We're just *finding out* how much we don't know. How does an eggshell get such amazing properties—its perfect shape, its self-assembly, its structural efficiency, its porosity to molecular oxygen so the embryo inside doesn't suffocate? And that's just one random example out of millions of possible ones. Everywhere we look there are mysteries.

"Then there's engineering. There's a whole tribe of engineers out there saying they're going to make nanoscale electronic circuits and machinery. But what if circuits and machinery aren't the *way* to compute on the nanoscale? What if simply keeping existing components and architecture, and trying to shrink them a few thousand diameters, won't work? Ever? Or what if it can be made to work in some clunky, wheezing fashion, but there's a much simpler way?

"No, no, no. We need to stop all the we-are-as-gods stuff. We need to prepare, experiment, observe, and learn."

This desirable approach to nanoscience, Perovic tells me, is called biomimicry. By following it, we need not reinvent the wheel—or the nanoassembler, the molecular catalyst, or anything else that life has already made workable in the last three thousand millennia. Microscale rotational motors, for example, already exist in nature. Among other things they power the whiplike flagella of the common intestinal bacterium *E. coli*. Another molecular motor, called F1-ATPase, rotates at a steady 800 rpm, the speed of an idling car engine, and measures just 8 x 14 nanometers. Human nano-engineers don't need to reinvent such marvelous biochemical devices, says Perovic. Life already provides us with working models. Many of these could be adopted for human needs virtually as-is. Others could require only analysis and tinkering to become synthetic machinery that "wet nanotech," bio-nano, could mass-produce. But in any case, as Perovic counsels, the best thing we can do right now is stop crowing about our knowledge and achievements, especially the ones we haven't made yet, and buckle down to work. There's lots of learning to do before we can start to design even the most elementary nanomechanisms.

Perovic has just established a working relationship with Toronto's Mount Sinai Hospital. He wants to extend this into a Bio-Nano Institute that will let humanity better understand the intricacies of biological systems.

"It's a wild and crazy time," he says, switching on a sunlike smile. "It's like the 1930s, when quantum science began generating atomic technology. This time we want to know: How do cells handle all those atoms they contain? What do intracellular atoms *do* inside cells? How are the atoms transported, delivered, received, or signed off? What are the biological rules for, say, electron ground-states?"

These questions aren't merely academic, Perovic maintains. They're literally vital to new technologies. Thin films and quantum dots, for example. "Qdots" are nano-sized bits of material that generate new wavelengths via fluorescence or stimulated emission. Says Perovic: "A few years ago, gallium nitride and aluminum nitride rocked the semiconductor world. Now we can make various sizes of light source from strips of those substances. What if we can also make sources that are tunable in output frequency and can be layered on the nanoscale? Then we'd have broad-spectrum white light out of a strip four or five atoms thick. That's a universally useful lightbulb with practically zero thickness, thousands of times thinner than a coat of paint."

In the near term, Perovic says, "Silicon in electronics won't disappear anytime soon. So whatever nanotech accomplishes will be most immediately useful and profitable if it can be integrated with silicon. Right now we're looking at single qdots twenty nanometers in diameter. In collaboration with [Canada's] National Research Council, we want to isolate them at the apex of a tiny pyramid and do spectroscopic analysis on them." Perovic says this single-molecule work would be an interim stage leading to spectroscopy of single atoms. To date, spectroscopy has been like polling a large population. Nanospectroscopy will be like interviewing individuals.

In the intermediate term, he adds, there's that whole notion of nanocomputing being different in kind as well as in scale. We really have to explore that, he tells me, rather than simply assuming we can go on doing what we've done so far and applying existing designs at smaller and smaller scales.

"A new microchip these days is only one centimeter square, but it's incredibly complex. Within that one square inch there may be several miles of circuitry. That generates a fantastically high heat density. Not only does it waste power, but you have to get rid of that heat somehow. Also, electrons aren't nearly as fast as light, which slows computing even at the nanoscale. One of our groups at U of T is working on an optical CPU [central processing unit]. They don't make any mention of dedicated circuits and waveguides. Once you're using light, they say, whether the photons are visible frequencies or microwave RF [radio frequency] or whatever, you don't need circuits at all. You can steer light using diffraction effects, or just broadcast it as a wavefront. Nanoscale CPU chips without circuitry as we know it—wouldn't that be the ultimate wireless technology?

"Then there's the concept of photonic paper—a thin film of polymers with properties that make them rewritable and erasable. We're very, very close to this material now. Give me a call about half-past 2003.

"In biological investigations, we're using a concept called beam blanking to greatly reduce the intensity of our probes. We're getting the probe beams so gentle, so low-power, that we can look at living tissue without disturbing it. For finer-scale investigations, we have a way of freezing soft tissue in situ so we can image it down to nanometer resolution in a high-resolution transmission electron microscope." What does he mean by high resolution? "A tenth of an angstrom. One percent of a nanometer."

Any thoughts on what's to come in the longer term—say, twenty years? Perovic drops his chin to his chest and considers. My, but this man can brood.

"We've come to understand somewhat how nanoscale forces scale up to our world," he says at last. "But to know what happens at the nanoscale itself, we may have to go beyond the nanoscale, to the subatomic level. After all, it's not just our world that's built on something smaller. The nanoscale must be, too." Good God! I yelp. The *picocosm?* Perovic laughs out loud. "If you want to call it that. Or the femtocosm, or the attocosm! That's where we're going. Nanotech is only the start, my friend. We're evolving our way to a revolution."

BIG ON MOLLUSKS

Dr. Rizhi Wang leans forward and passes me a sheaf of photomicrographs across his desk. I should say photo*nano*graphs. In some of them individual atoms stand out as clear and large as walnuts. We're examining the shells of bivalves, mollusks, in nanostructural detail.

"These features are on the order of a hundred angstroms," Dr. Wang says, tapping one of the images with a pen. Originally based at Tsinghua University in China, he's been a visiting professor and research associate at various North American universities for the past two years. His specialty is understanding—or trying to understand—the mechanical properties of substances at the nanoscale.

"This visual shows the mollusk's exoskeleton. It's rich in calcium carbonate, $CaCO_3$. When a few quadrillion of these creatures settle to the seabed over millions of years, they create thick beds of limestone."

Dr. Wang wants to know why such "biomaterials," materials produced by living organisms, behave as they do at the macroscale. Why is bone so strong and light? And in the present instance, why are mollusk shells so resistant to shattering, shock, and puncture? What traits do they possess at the nanoscale that make them so incredibly tough?

Once we know this, Dr. Wang says, we can duplicate biomaterials artificially. New products—scratch-tolerant kitchen counters, bulletproof armor for the police and military, energy-saving and high-power jet engines—await only this knowledge to be conceived and born.

Dr. Wang pries out his knowledge by every means he can. His methods include chemical analysis, X-ray, FTIR (fast Fourier-Transform Infrared Spectroscopy), and every form of microscope. He's also intimately familiar with the work of field-leading young scientists such as Dr. Joanna Aisenberg at Bell Laboratories and Dr. Angela Belcher of MIT and the University of Texas. Belcher was still a student at UCSB when she showed how mollusks use specialized proteins to form nanoscale $CaCO_3$ tiles. These tiles give mollusk shells their incredible resistance to shattering and puncture.

Before we get to the detailed methodology and findings of Dr. Wang's work, I ask a question that's begun to nag at me. Nanoscience is so new that nobody now in the field began in it. Everybody came from somewhere else—physics, chemistry, microscopy, and the biosciences. What got *him* into nanosci?

Dr. Wang considers the question a minute, then: "When I worked at Princeton," he tells me, "I found myself wondering why the fracture surfaces on a mouse's femur look the way they do when a mouse gets a broken leg. Why do some of the newly exposed surfaces have a zigzag profile, while others are smooth? Parallel to the bone's long axis, the bone usually splits clean and straight. In that direction the bone has perfect cleavage planes. But when the break goes *across* the femur, at right angles to the long axis, the broken surfaces are as serrated as a mountain range. Why?

"I began to think it had something to do with the structure of the bone, [but] not the microstructure; that was well documented, and nothing in it gave a complete and satisfying explanation. I realized I'd have to go deeper, down into the nanoscale makeup of the material."

Bone, he explains, is a nanocomposite material made by nature. It comprises a protein matrix, throughout which nanoscale particles of mineral are embedded. The matrix is mostly collagen, a tough fibrous substance that the body uses to knit skin. But in bone, down at the hundred-nanometer scale, the collagen contains nanocrystals of a calcium compound. These are not merely associated with the collagen fibers, or suspended in them. The nanoparticles and the matrix are strongly held with a tight chemical bond. "We think there are protein molecules of low molecular weight, around which the nanoparticles nucleate," he says. I give him my trademark stare: Dumb it down, please. What does he mean by nucleation? He backs up and tries again.

"A snowflake or a raindrop needs a tiny particle of airborne dust around which it can crystallize," he says. "The particle triggers the crystal, so to speak. In a similar way, we suspect these small proteins that we find in bone at the nanoscale jump-start the formation of the small calcium particles within the bone. The same proteins may also be the glue that knits the nanoparticles to the bone's collagen. On the microscale, bone is nothing more than a grouping of these composite fibers." Clearer. But how do you *know* all this? That gets me an enthusiastic smile. *Ah! Methodology!*

"There were some elegant experiments. Some of them were transgenic technologies that disabled key parts of mouse DNA. This led to tiny changes at the bone's nanoscale, which in turn affected the large-scale properties of the bone. Simply changing how the matrix fibers were coiled created bone that was far weaker and more breakable. That came from deactivating a single gene."

In bone, Dr. Wang goes on to say, nature greatly varies the ratio of calcium (the hard stuff) to collagen (the soft stuff). At one extreme there's the rostral bone of a toothed whale, with a 6:1 ratio for hard to soft material. It's three times as stiff as normal bone. Says Dr. Wang: "That type of bone is brittle, but it doesn't matter. The animal doesn't need it to be strong. It's used to transmit sound, letting the fish sense prey or danger in murky water. Its high density makes it extremely efficient at its task."

Structural bones, on the other hand, including human bones, usually have a 3:2 ratio. That makes them slightly resilient yet sufficiently hard—in other words, tough.

A natural material even stiffer than mammalian bone is the outer shell of the common mollusk. Its nanostructure, Dr. Wang explains, is the mineral aragonite, a crystalline form of calcium carbonate. The nanostructure of the mollusk shell contains countless plates of this substance. "Each plate has precisely engineered nanoscale features, and vast numbers of plates are packed like stacks of cards. On the micron scale, this creates a structure like a brick wall.

"You can look at the building, or you can look at the brick," Dr. Wang says. "I look at the brick—that is, material structure at the finest possible scale. How, I ask, does that nanostructure affect the mechanical properties I see at the macroscale and the mesoscale?"

The mesoscale is the world between the nano and the macro; its features are on the order of 0.1–100 microns, or 100–100,000 nm. At this scale, the mollusk shell is incredibly strong. A close look at the mesoscale-nanoscale interface reveals why. The mollusk shell has a unique structure. Its plates of natural nanocomposite are so tough because they resist sliding across, or separating from, adjacent plates. Each plate self-assembles on its surface a series of trapezoidal bumps, shaped like flattened pyramids. In effect, this dovetails adjacent plates together: The bumps fetch up against each other and prevent sliding. Even when there's a slip in response to building stress, the nanostructure falls back to a new defensive position slightly downfield and digs in just as strongly (see Figure 2-1). If forces are high enough to shear off or override one bump, no matter. The next bump will repeat the process, again dissipating energy and resisting

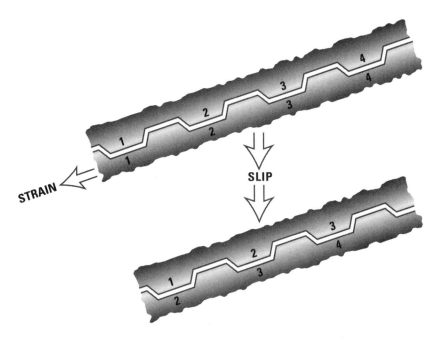

Figure 2-1. Relocking of mollusk shell fibers

any slide. It's a relentless defense that never cries uncle. The plates do this again and again, grudging every angstrom of lost ground. The net effect is to isolate any dislocations and delaminations that stress does manage to induce in the shell material. To the last gasp, mollusk shell keeps these flaws from joining up and creating one big structure-threatening crack. In technical terms, nacre—as bivalve shell material is called—works like a ceramic that, at the nanoscale, can flow slightly. This trait, called plastic deformation, is what makes steel so tough.

Compare this tenacity to standard ceramics, whose nanostructure does not block crack propagation. Ceramics, like stone, lack the ability to deform plastically at the nanoscale. This nanoproperty has immediate consequences in our everyday world. Without warning, an almost invisible hairline flaw in stoneware can flash from the nanoscale through the mesoscale to the macroscale. Many of us learn to avoid cracked china only the hard way, after great-grandmother's Spode dumps hot tea in our laps. Now you don't see the cracks, now you do.

By contrast, the nanoscale crack-resistance of mollusk shell translates into extreme toughness at both the mesoscale *and* the macroscale. Again in technical terms, the shell is "notch-insensitive." The fine cracks and dislocations that precede snapping or shattering simply cannot propagate through mollusk shell without huge, almost extreme, forces coming into play. Sea otters have to float on their backs and repeatedly smack mollusk shells with large rocks to open them. Ravens must take mollusks in their beaks, fly into the air, and drop them on hard surfaces from fifty feet up to crack the shells. Moreover, these two animals are among the brainiest in existence. Less intelligent critters, gulls for instance, are utterly and permanently baffled by the strength of mollusk shell. Tons of food at hand for the taking, cached in small and defenseless units, and all unobtainable. Five million years of evolution haven't given the average gull the sense to crack a mollusk. Seafood, seafood everywhere, nor any bit to bite. Score one for the defense.

Dr. Wang's work exhibits a trait found in all the best science. Years of careful experiments, done solely out of curiosity, suddenly point the way to practical applications. In Dr. Wang's case, it's the possibility of bone implants as strong as the bones of a healthy twenty-year-old athlete. In ten years or so, such implants may even be grown in place without surgery. Rizhi Wang doesn't think it's impossible.

"We're establishing a joint research program with some teaching hospitals," he says. "But it's not just medical applications that will come out of this work. We're on the trail of synthetic ceramics that are lighter than aluminum, more heat-resistant than firebrick, and tougher than steel. Imagine car engines one-third their present weight, with twice the fuel economy. Once we have the knowledge, there's no reason why it can't be done."

How long before this happens? He thinks, shrugs, spreads his hands, and smiles. "I'm no prophet." But? A bigger smile. "Within six years."

COCKTAILS AND COMPUTERS

Ever since the English cybernetics genius Alan Turing suggested it half a century ago, the thought experiment named after him has been a touchstone for anyone working in the rough backcountry of artificial intelligence. Humans will never accept a computer as intelligent, Turing said, until we have a conversation with it—not necessarily verbal, typing would do—and are unable to tell whether we were conversing with a fellow human or with a machine. At that point we would *have* to accept machine intelligence. It would be upon us whether we liked it or not.

Much has been written about the Turing test, but one fact has emerged with awful clarity. Just as it was hubris on the part of science to dismiss nature's material achievements, so it was (and is) a mark of our own willful human stupidity to diss our own human intelligence. It's a striking measure of how much we've learned in the last two decades that few reputable scientists in any field now make glib predictions about robot citizens with positronic brains. As much data as we churn out on the human brain—its powers of mind, memory, association, recognition, linear reason, parallel processing, and mad intuitive leaps—still our own brain rises above us, as one despairing critic said of Bach's music, like a summitless mountain. We're not there yet, folks, and we may never be.

I once discussed the Turing test with Dr. Tom Theis. Thomas N. Theis (rhymes with "nice," a good description) is director of physical sciences at IBM's Thomas J. Watson Research Center in upstate New York. After sitting together for an hour one morning patiently listening to a bizarre Drexlerian sermon in Silicon Valley, we started whispering like a pair of naughty schoolkids about what directions respectable nanotech might really take. Tom was full of fabulous ideas. A ten-minute chat with him cost me ninety minutes in frenzied note-taking afterward, recovering dialogue from memory and adding follow-up queries of my own. But at one point in our talk, I did manage to make his jaw sag—whether with awe, pity, or incomprehension I don't know.

"Maybe," I suggested, "a computer won't pass the Turing test until it can make you feel guilty."

I tossed this out frivolously, but the more I chew that thought the more it cracks my teeth. Maybe there's a core of truth in a remark that I meant as merely flippant. Maybe it won't be enough to put a data-processing machine through its paces in data processing. *Name the capital of Malaysia*, or *Give me the cube root of 2*, or *Define "astronomical unit"* all have sound, unambiguous answers that a machine could present as easily as a person. More easily, in fact. It might behoove a machine to fake ignorance now and then, lest it blow its cover. ("Jeez, I dunno. Whaddaya think I am, a freakin' computer?")

That's not enough to let a machine pass a Turing test. We must, in communicating with an intelligent machine, sense emotion: not just guilt but passion, anger, uncertainty, and love. Furthermore, we must sense that emotion *because it is there.* If some current theorists are right, and I suspect they are, emotions are quick estimates of critical data states. These are processed lightning-fast because the brain-owner's continued survival may depend on such high speed.

Many academics do poorly in business because they've overdeveloped their linear processing skills at the expense of the more quick-and-dirty methods of data processing. This includes intuition, which makes itself known via sudden elation or unease. Yell *Fire!* in a theater and most patrons leap for the door. Those who calmly proceed to strike an *ad hoc* committee to study the problem are probably professors.

There's certainly no doubt about their species. Any being prone to steady, thorough pondering is *Homo sapiens,* no mistake. Dr. Roger Fouts is director of the Chimpanzee and Human Communication Institute at Central Washington University in Everett, Washington. He believes the brains of primates—that's us and the great apes—process ultrafast reactions and sequential thought with subtypes of neural tissue that are different and distinct. *Gray* matter, which dominates humans' frontal lobes, takes care of linear thought. *White* matter handles reflexes and intuition. Chimpanzees' brains have a far higher proportion of white matter to gray than we do. This small variation in our morphology, attributable perhaps to one or two genes, defines much of our mental difference. Our two species' genomes overlap by nearly 98 percent. A human is just a chimp with a couple of dealer trim options. Fouts titled his book on human-chimpanzee dealings *Next of Kin.*

All this suggests that however many transistors we may one day cram into a cubic angstrom, we will never construct machine intelligence until we can make hardware that mimics the brain's white matter as well as its gray. A workable AI may not make you feel guilty, but it should—it must—display and evoke a full range of other emotions in anyone with whom it talks. Fanciful? A common fact shows it's not. All over the world since the invention of e-mail, couples separated by yards or miles regularly fall in love. If communication is mere words on a screen, lines bent into letters to encode speech phonemes and through them linear thought, how is this proven effect possible? No exchange of pheromones or fluids. No heady perfume, gleam of teeth, oiled pecs, leers over martinis, or heaving alabaster bosoms: just the cipher called language. Yet bonding hormones arise as surely as if the mate of our dreams had walked into the room and given us the eye. What's going on?

It's this, I think: the two humans in an e-mail affair have taken and passed their own Turing test. Somehow, God knows how, cues—tiny quanta of meaning embedded in word or usage, or that amazing and unquantifiable element called humor—have been sent, received, and responded to. And sight unseen, skin unfelt, glands unscented, attraction takes wing. I'm short on concrete predictions, but here's one. It's going to be decades before any machine can flirt. I won't say never; I do say many years.

Another property of true intelligence, one even less heralded than emotion, is its ability to discriminate. *It's a good wine but not a great wine. This whiskey is fine; that one is plonk. Monet the genius but Manet the hack,* and so on. We call this discriminatory power "taste" because like that sense's ability to discriminate among complex molecules, it is both keen and educable. And surprise surprise, another trait at the core of intelligence is an ability to adjust perception and behavior according to experience—in other words, to educate itself.

Everyone knows about standard learning, the buy-low, sell-high kind, or the reflex that prompts you to squawk when your finger hits a stove. But an active brain learns incessantly. Even at night, it learns. One function of dreams is to winnow stored information, keeping what's important and filing away the rest.

Another type of taste discrimination, which in humans is both educable and almost universal, drives acoustic engineers absolutely barmy with incomprehension. It's called the cocktail party effect. While nearly everyone has experienced it, it still hasn't been properly explained. You're at a party—not a rave that's louder than a cymbal factory; say 60 dB, a comfortable hum—and cruising for a group to join. You don't mind talking politics, but you're not about to leap into a discussion about whether alfalfa sprouts have souls. You stand equidistant from three groups and eavesdrop. This is what you hear:

"... each other. So why *shouldn't* a ménage à trois work? I mean ..."

"No. No. No. No. No, no, no, no, no, no, *no*. No. No. Nonononononono."

"I hear you, Jim, but you have to look at the other side. If Bush ..."

Group One, right? Absolutely. But what's most fascinating about this scenario is not the various styles, contents, or social overtones of the talk. It's the easy ability of the listener's brain to tune into a series of equally loud speakers, one after the other, at will. This is the cocktail party effect. And no one has yet figured out how the human brain (especially in one trained to discriminate, such as a socialite hostess or a professional spy) accomplishes this everyday feat. The more you analyze it scientifically, the more the cocktail party effect seems like witchcraft. Somehow, the listening brain identifies a set of variables associated with the one voice it wants to follow—tone, range, frequency, or a combination of these and other things—and accepts cognitive data from only that source. It can't work; it shouldn't work. But every day it does.

Nanoscience may be on the trail of this fascinating bit of biomimicry, which has withstood inquiry for so long. If basic research now under way does ferret out the algorithms behind the cocktail party effect, it should quickly find its way into commercialized technology. The application is not surprising; the research approach is.

Several million people around the world wear a hearing aid at least part of the time. However small its electronics may be, the device itself is conceptually simple. It's a broad-spectrum magnifier of sound-wave amplitude—a miniature PA system comprising mike, preamp, amplifier, and speaker. It listens to what's going on around you, then relays it to your outer ear in a bellow. Some hearing aids amplify some frequencies more than others, like the simple tone dials that radios possessed before graphic equalizers were commonplace. But on the whole, hearing aids do not discriminate. They yell what they hear, whether it be a whisper in a still room, a dog barking, or the roar of a nearby jet.

At a party, those who wear hearing aids experience a wash of sound as the hum of many voices blends into a running river of white noise. A hearing aid with enough AI to use the cocktail party effect would be a godsend.

But how and where to start? Dr. Simon Haykin, a visiting professor at Purdue University in Lafayette, Indiana, suggests we begin at the deep conceptual level. Dr. Haykin, an electrical engineer, is an expert in adaptive systems—of which, he tells me, the human brain is the best example we know.

Dr. H is a fascinating fellow—white-haired and avuncular, dressed in a moth-eaten sweater and slacks. He's a ringer for Bilbo Baggins at age

105: hale, hearty, and energetic. He has a vast new-kindled excitement for nanoscience. I meet him in a big cluttered office that looks out onto massive construction. In an adjoining anteroom lurks the definitive senior-prof secretary, tinier even than Dr. H, taciturn and acerbic.

Dr. H pumps my hand, motions me to a seat, and launches at once into his *recitativo.* "Computers today all use the initial von Neumann paradigm," he says. "That means, first, that they typically are digital. They use binary algebra, a series of circuits that are either on or off: nothing in between. The second part of the von Neumann paradigm involves serial computation. So today's computers are serial: They chew through problems one bite at a time. Usually this architecture is not very elegant; in fact, it's a brute-force method that solves complex problems only by virtue of its extreme speed.

"Unfortunately, our computers seem to be near the limit of this approach. Their hardware is already pushing the maximum possible speed for electrons moving through a solid. Theory says that, even given that speed limit, you can reduce processing time by packing chip components more and more tightly together, minimizing the electrons' path lengths. But the transistors and other elements in an advanced CPU chip are now so crowded that they create heat on the order of ten watts per square centimeter, which must be got rid of somehow. And still the total circuit length on a modern chip is over a mile!

"It's obvious to me that miniaturization, as a route to hardware innovation, is a dead-end street. It's given us some noteworthy achievements, but now it's run smack up against its ultimate limitations."

Dr. H rises from his battered wooden swivel-chair and draws a seventeen-cell matrix on his blackboard, explaining as he goes. "Conventional computers use logic gates. These operate in the nanosecond range and drain one microjoule per operation per second. The brain's equivalent to the logic gate is the nerve cell, or neuron. It's a million times slower and burns ten billion times less power than a microtransistor logic gate. But it works, because it's organized differently. Its computing architecture is not serial; it's parallel—massively parallel. Great quantities of neurons chew through problems from many directions at once, then assemble those separate findings to get their final result. And guess what? The brain performs a lot of high-order operations far better than the most advanced machine we have today. A supercomputer takes days to analyze a human face and classify it as Unknown or else attach a name to it. The human brain completes the same task in between one-tenth and one-twentieth of a second. It also analyzes more complex data sets to reach really subtle conclusions such as age, gender, and emotional state. That's beyond today's computers altogether." Dr. H flips his chalk in the air. "I think it's obvious what direction we should go in."

Tomorrow's nanocomputer, Dr. H explains—the kind you'd need to give a hearing aid the cocktail party effect—would be massively parallel. It would break big tasks into little tasks, which it would then process all at once rather than one after the other. To minimize heat-dissipation problems, it would cover a large area. Dr. H foresees something wearable—call it CompuCloth—whose circuits are lightweight, rugged, and mechanically flexible. It would be produced by printing onto film or fabric, using techniques like silkscreen or batik. Because of this, once CompuCloth's R&D costs had been written down it would be extremely inexpensive to produce in both high quality and high quantity. These wearable polymer electronics would represent another form of biomimicry. They would reproduce not the shape or chemistry of the brain, but its functionality—how it works.

CompuCloth (ComBatik?) could lead to a lot more products than intelligent hearing aids. How about musical instruments you wear, with an entire brass section that fits into a wallet? How about an advanced graphics computer the size, shape, and weight of a pocket handkerchief, that you can carry as conveniently as one? (Social etiquette note to future computer users: Examine what you blow your nose in before the fact.)

"The interesting thing about all this," Dr. H says, returning to plunk himself down in his chair, "is that von Neumann himself advocated it just before he died. If he'd lived, perhaps we'd be calling this later approach 'the von Neumann paradigm.' It's certainly the paradigm of the man's own brain—and yours, and mine, and everyone's."

But, I object, wouldn't almost anything—a mosquito biting through the computer-vest, for instance—interdict its function? Dr. H grins. Obviously he's had this question before, and he's ready to field it. He tells me about defect-tolerant computing, another aspect of his proposed fabric-electronic architecture.

"The Teramac," he says, "is a massively parallel experimental computer that Hewlett-Packard Laboratories in Palo Alto designed and constructed in 1994–1995. Its designers built in over two hundred thousand hardware defects, any one of which would have been instantly fatal in a computer using conventional serial computation. Yet the Teramac consistently operates two orders of magnitude faster than the fastest single-processor commercial workstation."

The Teramac, it turns out, has enough knowledge (i.e., self-knowledge) to find and use paths and components alternative to the damaged ones. The experimental H-P computer is like a driver who knows her neighborhood so well she can instantly nip into a laneway to avoid a looming traffic block on the main thoroughfare. Again, says Dr. H, this is how our own brains work.

"You can view neurons as unreliable components if you like. And individually, they are. They're rather messy things. But that doesn't matter, because there are so many of them. More to the point, there are so many possible paths that link them. The possible number of connections among neurons is staggering."

Dr. H ain't just a-woofin'. The mathematical expression for all possible links (L) among X discrete elements is X!, pronounced "X-factorial." Thus:

$$L = X! = X(X-1)(X-2) \ldots 1$$

In other words, the possible connections among 10 elements is:

$$10 \times 9 \times 8 \times 7 \times 6 \times 5 \times 4 \times 3 \times 2 \times 1 = 3{,}628{,}800$$

I'll leave you to calculate the connections among 10,000,000,000 cerebral neurons. Dr. H suspects the result exceeds the number of subatomic particles in the observable universe.

"Two things are apparent," Dr. H says. "First, we can learn a lot about nanoscience and its technological applications by studying the natural nanosystem of the human brain. Second, while everyone agrees that nanotech requires a lot of different disciplines to make it work, most people tend to list only physics, chemistry, biology, and materials science. Who'd have thought that electrical engineering would also come in handy?"

Dr. H has been traveling around the world examining nanotechnology before assembling a teaching and research program at his home university, McMaster. "I'm amazed at what's starting to happen," he tells me. "After health and defense, this is the third-highest level of research funding in the USA."

Where would he see the big discoveries coming? Dr. H holds up his hand and ticks off fingers. "The role of polymers in nanosystems. Bio-interfaces. Optoelectronics. Adaptive systems. Those are all going to be immense."

If he looked forward ten or fifteen years, what would he see? "There's a teaching hospital I want to get involved with," he said. "I won't tell you where. But they're starting up a Brain-Body Institute to look at how mind and body interact.

"Think of it! If you organize scans of the entire body using PET and NMR, and then analyze those medical images, you could be able to explore how brain and body communicate. You could see how stress relates to illnesses such as depression. You could isolate the chemicals that bring this about, and follow them throughout the body as they move and change.

"It's a very, very exciting time in science," Dr. H concludes. "I'm glad I've lived to see it."

NEW THRESHOLD, NEW LAWS

There's one thing I should understand straight off, says Dr. Donald Sprung: "I'm a theorist. My work is mathematical or computational; I haven't done any experiments since I got my Ph.D."

That life has been long and busy. At McMaster University, a small school with a big research profile, Sprung has been both dean of science and chair of the Department of Physics and Astronomy. His business card lists him as professor emeritus of physics; his degrees and affiliations are Ph.D., D.Sc., and F.R.S.C. That last one means Fellow of the Royal Society of Canada, an honor reserved for major contributors to science north of the 49th parallel.

In appearance, Sprung seems to have teleported straight from the 1950s. He's tall, ramrod-straight, and clean-shaven, with a shock of snow-white hair. He's also dressed impeccably in slacks, shirt, tie, and sports coat. I had thought that only odious people, dull civil servants and sneering young movie stars, wore sports coats anymore. To my mental list I now add elegant professors emeritus of physics, who are admirable rather than odious.

Sprung and Simon Haykin reinforce something I've begun to note: the quantity of gray hair I've found in my research into the nanocosm. While nano*technology* is dominated by men and women in their thirties and forties, a lot of nano*science* seems the province of grand old men. Social critics excoriate our culture as youth-obsessed, but here's a striking exception. Perhaps nanoscience is such a newborn that its umbilicals to the mother disciplines have not been cut. Young fields may favor old practitioners.

However aged these nanoscientists' bodies, there's nothing geriatric about their minds. They are anything but stodgy or pedantic. Their ideas are as radical as anyone's, and they're full of missionary zeal. *When the spirit is proud,* wrote the Greek author Nikos Kazantzakis, *it stands erect and does not permit the years to touch it.* Of course, my sympathy may come from a recent realization that I need bifocals, too.

Whatever the cause, as Sprung talks I'm taking notes in a frenzy of speed to keep up with what he's saying. It's from unlikely places and people like this, a semi-retired old man in a small university, that transforming concepts often come. No place on earth has a monopoly on ideas.

Sprung draws diagrams as he talks, as if he's lecturing undergrads at a blackboard. This makes it hard to pay attention to his words, because I'm so struck by his manner. He speaks so quietly he's almost whispering; to make charts and diagrams, he moves his mechanical pencil so lightly that its lines are little darker than the paper they're on. Yet despite being almost inaudible and nearly invisible, Sprung has an air of calm authority that makes me want to doff my cap and say, "Ess, Sorr." I realize his tentativeness has nothing to do with shyness or uncertainty. Professor Sprung is merely a scientist in the classically precise mold. He would never dream of saying "that cow is brown." He'd think awhile, then murmur: "This side of that animal is apparently dark, subject to independent scientific verification." I'm not slagging the man; it's just that he evokes awe. When he talks, it's as if nature had taken human form to explain why the gravitational constant has its unique value.

"Below fifty nanometers or so," Sprung tells me, "the individual electron can sense that it's confined. At these scales, it stops behaving like a charged particle and starts behaving like a wave-function. What is it really? Well, it's both. Physicists call it a 'wavicle.' We label it, and treat it, according to its behavior in any situation.

"It's unfortunate, but some people trying to design molecule-sized circuits continue to visualize electrons as little elastic billiard balls instead of as waves. That creates nonfunctional circuit designs at the nanoscale.

So, how do you do things properly? You do what you always do when change of size reaches a threshold that also alters properties. You adapt."

I ask him for an example. "An example? Let me see . . . Yes. Any notion of 'electric current' has to be abandoned when you design mesoscale circuits. A microchip deals with electron quantities so small—nanoamperes in some cases—that electrons must be treated as a discrete stream of particles, rather than as a flowing fluid. The metaphor must change to reflect nature's changing behavior."

Sprung's tentative pencil sketches a corridor with several dead-end niches opening off its walls. "A phase transformation occurs at the border between the mesoscale and the nanoscale. Almost by definition, a nanoscale circuit has room for only one electron at a time. I've drawn a circuit path about twenty nanometers wide. A particulate electron would bounce *here* and *here*. Then it would return to its start point."

I recognize the effect. It's a corner reflector, which returns incident light directly back to source. If you look closely at a bike reflector you'll see an array of 500-micron cubes molded into Lucite. These cubes are placed corner-on to the plane of the reflector. They ensure that anyone driving a car will see his headlamp light shot back to him.

This thought takes me 100 milliseconds, which is enough to lose the thread of Sprung's discussion. I'm sweating now, as if I were running a hard race. Sprung continues, as calm, unhurrying, and inexorable as a glacier.

"Here's how the nanoscale electron—a single electron—really behaves. It stops at the entrance to a niche and holds there, behaving as a standing wave. If you put a voltage at the closed end of the niche, you squeeze the wave out and the electron goes on its way. That lets the corridor circuit function as an elementary logic gate: a nanotransistor." *Nanosistor,* I think, and lose another hundred milliseconds. This time I cover my tracks by interrupting him:

"Sir [*Sorr!*]—you're a theorist. How did you make this discovery if you never ran an experiment?"

Sprung looks at me in mild surprise. "I calculated it, of course. My calculations were afterward duplicated by some of my colleagues."

"Did someone verify this experimentally?"

"To a degree," he says. "The idea is accurate. It turns out, however, that for this design you need to input a great deal of information for each—"

"Nanosistor?"

"—each finite element," Sprung continues, lifting an eyebrow as he ignores my churlish attempt at an interruption. "You must calibrate each element, each cell. Hence to have a linear array of several million of them, as required in any workable nanochip, would be prohibitively expensive."

"So your calculations were a failure?"

Sprung looks up at me, a mild half-smile at the corners of his mouth.

"Not a failure, no. It's a cost-effective way to develop technology if you can show that a promising option is really a blind alley. It's even better to do this before you run an experiment, and best of all to do it by calculation alone."

The old man gazes contentedly out the window. "Besides, it did involve some elegant mathematics. Oh my, yes."

NEW LIGHT ON SOLAR POWER

Dr. George Sawatzky, another mature scientist staking out new ground in the nanocosm, was described to me by several sources as one of Europe's foremost thinkers. Originally of Russian descent, he has been a full professor of the University of Groningen, the Netherlands, for the last thirty years. I mentally gird myself for an accent as thick as mayonnaise—and to my surprise, hear him speak in flawless English. Evidently, this is a man of many skills.

Dr. Sawatzky, I learn, got his Ph.D. in Manitoba before settling in the Netherlands as a professor of physics and physical chemistry. His specialties include inorganic materials that are magnetic and superconducting—that is, they are able to transmit electric current with little or no resistance. In pursuing this work, Dr. Sawatzky recognized early on that all physical properties of materials strongly depend on their molecular nanostructure. The trouble was (and is) that the properties of a macromaterial, the substance we deal with in our everyday world of jam jars and jet engines, is one thing; its nanostructure is another. So far nobody's come up with a way to link the two. It's still impossible to forecast from a crystalline nanostructure how a macromaterial will behave. Drexlerians, take note: This is why experiment must test, and then extend or refute, all theory. To think in a vacuum risks producing theoretical structures of unrivaled elegance that are also dead wrong.

Macroproperties, it turns out, are impossibly complex. Nature may spring from a handful of elementary particles; but even by the time the subatomic reaches the nanoscale, these few varieties of matter have combined and recombined to create structures, substances, and energy fields so involved that they defy prediction and ultimately may defy even comprehension. Nonetheless, thinkers like George Sawatzky still struggle to wrest from nature as much understanding as she will allow.

Dr. Sawatzky is blunt about his work: "It's nanoscience, not nanotechnology. I deal with basic knowledge. But if science has shown us one single thing, it is that if you do science properly, you create a strong foundation for later applications. Technology follows rapidly, almost automatically it seems in retrospect, from good scientific work."

As Dr. Sawatzky takes me through his recent research, I see several improved technologies that his studies could support. Practical solar cells, for example. The photoelectric effect, by which light falling on certain materials creates a flow of electrons, has been known for over a century. Albert Einstein based his doctoral dissertation on it. But "active solar," as greenies call the large-scale conversion of sunshine to grid current, has grown only slowly. There's one big reason for this. Developing solar-cell materials that are rugged, long-lasting, cost-effective, easy to produce, and efficient at energy conversion has proved to be an absolute technological bitch. The thermoplastic polymers in which they are embedded degrade from transparency to translucency because of ultraviolet cross-linkage, the same effect that fades your curtains and embrittles your kids' plastic toys. There's also the matter of particulate abrasion:

Q: Where's the best place to put an active-solar panel?
A: Where there's the most sunlight.

Q: And where is that?
A: Where there's minimal cloud cover.

Q: Which is where?
A: The driest areas—deserts.

Q: And what are deserts full of?
A: Airborne dirt and grit.

Q: How does that effect solar panels?
A: Like a high-power sandblaster.

Obviously, those who would get earth's power from sunlight rather than from fossil fuels must go back to the drawing boards. Solar's not here yet. But work persists, because it's so tempting a dream. Humanity can have few greater goals than everlasting energy free for the taking. Ten kilograms of pure energy have fallen on our planet every day, year in year out, since earth accreted four billion years ago. That constant allotment of solar power is four or five orders of magnitude more than our entire civilization needs. Some of the sun's input goes to weather; it evaporates water and moves the clouds around. The fury of a hurricane, the fire of lightning, are just the rapid output of solar power banked in the ocean and the clouds. In exactly the same way, a capacitor stores electrical potential and then zaps it out in a short, sharp spark.

Some solar input binds carbon dioxide and water into long-chain sugars called wood. Other sunshine drives the ocean currents, warms our swimming pools, and grows our gardens and lawns. But the vast bulk of it is dumped back into interplanetary space each night as re-radiated infrared—waste heat. If we could harness the tiniest fraction of this squandered power, we would at a stroke stop global warming. We could mothball our supertankers and refineries, cap our oil and gas wells, and tell Dick Cheney and the rest of the sludge barons to go earn an honest living. "Solar cells," Dr. Sawatzky points out, "emit no greenhouse gases. In fact, they create no waste product at all until they break down."

Did I say that achieving this goal requires a return to the drawing board? Sorry, slip of the pen. CAD/CAM is actually an advanced design stage. Long before we reach that point, we must go back to the experimentalists' labs and the theoreticians' equations. That's where Dr. Sawatzky comes in. He is perhaps the world's foremost expert in understanding what happens in the nanocosm when light and matter interact.

An organic solar cell, he explains, could be made in quantity if you first burrowed down to the nanoscale. You could then replace the heart of existing solar cells, called a P-N-P (or positive-negative-positive) junction, with a simpler P-N junction. Furthermore, you could do this in limitless quantity. You could design a synthetic molecule that traps incident solar energy, splits it into positive and negative charges, and then transports these charges to storage areas along a series of electron collector-roads. This solar battery would use two poles, anode and cathode. But it will be brought into existence only if materials with the right nanostructure can be found.

At least one material combination is already known that satisfies requirements for a solar battery of this type. It is a core of pure aluminum, coated with a carbon form called C_{60} and a polymer such as sexithiophene.

Unfortunately, the real world comes crashing into this elegant theory. Operating in air, the aluminum portion of this solar battery would quickly disappear beneath a ten-nanometer film of Al_2O_3. That's aluminum oxide, the stuff that dulls the finish on your patio chairs. In large quantity it's called corundum, which after diamond is the hardest substance known—hard enough to sharpen a tempered-steel knife. While quick to form on metallic aluminum, Al_2O_3 is difficult to get off; and so far at least, it's proven impossible to prevent from accumulating.

"Solar cells," Sawatzky tells me, "are made by coating a metal with an organic substance. But this is done in air, so that the metal will always be coated with a thin layer of metal oxide—a substance that is electrically insulating. We have to find ways of preventing this oxidation, or at least controlling it.

"Organic molecules also break down rapidly if they trap an electrical charge. Thus a solar cell in which charges can't move quickly and easily to the electrodes won't last very long. We need to discover what's going on where the cell's metal meets its organic wrap. And that region is less than one nanometer thick."

To find alternative solar-cell materials, or else to solve the problem of oxygen corrosion at its source, Dr Sawatzky is in his own words behaving "half like a theorist and half like an experimenter." He adds with a smile: "One of the most interesting questions I've posed myself is: 'How do I predict properties?' Nature is so wonderfully varied. Take sodium chloride, NaCl. In its most common solid form it's a perfectly cubic crystal. That's what you find in your table salt shaker; that's what you put on your food. When this crystal splits, it displays lovely flat cleavage planes. But when it's grown on a certain kind of substrate, it adjusts its crystalline form. Now it alternates atoms: Na-Cl-Na-Cl, et cetera. And this form of sodium chloride has entirely different properties! It has new characteristics that seemingly come out of nowhere and can be truly bizarre."

The new material he has theoretically predicted, Dr. Sawatzky explains, is a "forced crystal" which, on the nanoscale, duplicates some of the properties of its substrate. Among these properties is an ability to store vast amounts of electrical charge: in other words, to function as a nanoscale capacitor. This may be the long-sought key to limitless solar power—and a clean, prosperous society based on table salt.

An ideal organic material for a solar cell, Dr. Sawatzky says, would act like a two-lane highway, carrying electrons in one direction and positive charges (called "electron holes" or just holes) in a diametrically opposite direction. The solar material would also self-assemble. This should be possible, he says. Even schoolchildren have been growing simple crystals such as cupric sulfate (CuS) for years in aqueous solution.

"Even if a solar-cell crystal proves to be far more complex at the nanoscale than sodium chloride," he tells me, "we may be able to apply certain natural forces commonly found in biology. The ones, I mean, that in less than a second distort a chain of amino acids into a precisely shaped, functioning protein. The folding of a macromolecule is far more subtle than standard crystallization. It uses weak bonds among atoms that are not adjacent, but relatively far from one another."

Dr. Sawatzky also sees an adaptive application for the "channeling molecules" that transport ions across the membranes of living cells through gaps only a few nanometers in diameter.

"This could lead to nanoscale inkjet printers that assemble themselves," he tells me. "But that's probably a few years down the road."

Dr. Sawatzky's eyes shine as he paints his vision of how nanotechnology should develop. "It doesn't pay to take mature scientists and try to educate them in other disciplines," he says. "A chemist must stay a chemist, a physicist must stay a physicist." He laughs out loud: "It doesn't even matter if they talk to one another, as long as their graduate students do!"

George Sawatzky leans back in his chair, putting his big hands behind a shock of white hair as wild as his ideas. "Only intuition can tell a scientist what to look at next," he says. "I'll tell you one thing, though. Every discipline in existence, engineering and bioscience and chemistry and physics and whatnot, is on a collision course. Chemistry has been increasing the scope of its investigations, looking at larger and larger structures. At the same time, bioscience and engineering have been reducing the size of what they examine—from living cells down to large molecules. Do you know what this means?"

I shake my head.

"It means," he says, "that science will change forever. All these existing disciplines are about to run into one other at the nanoscale."

CHAPTER 3

NANOTECHNOLOGY
TRENDS in WORLD
DEVELOPMENT

Fool me once, shame on you
Fool me twice, shame on me
— Old saying

CASH AND RISK

FOR SOMETHING that's so new to serious investigation, the nanocosm has shown itself amenable to commercialization at a record-setting pace. Half a century elapsed between Faraday's experiments with "the Electrical Fluid" and the commercial supply of DC power to Paris, London, and New York. Penicillin took two decades to move from benchtop observation to prescription drug. But today, the typical journal paper on nanotechnology will already have taken the first steps to commercialization by the time it appears in print.

This striking situation is unique among present-day science and technology. Outside nanoscience and bio-pharmaceuticals, it's a bad time to ask business to support a technical venture. By mid-2002, the business climate was as hostile to new knowledge-based ventures as it was in the mid-1970s. Nanotech had to overcome some steep financial hurdles to achieve its present success. You need look only as far as the dot-com explosion to see why.

As with many disasters, the effects of the dot-coms' meltdown fell mainly on the innocent. By 1993, advanced technology had convinced most mainstream banks and venture capitalists that plans didn't need bricks, mortar, and land to be worth funding. Intellectual property (IP), investors realized, could be collateral as sound as real estate.

Then the investment pendulum swung too far the other way. As if to undo its years of underfunding and neglecting high-tech, capital went on a spending frenzy, throwing money at every kid with pimples and a technical degree. Dot-com mania became a Florida land bubble for the chalkstripe set, who often ignored due diligence in their lust to profit from high-tech. Yet it's obvious in hindsight, and should have been obvious at the time, that wild wishes are not IP; that sound practice, both in financing a business and in operating it, has no cheap substitute; that an IPO is not a business plan; and that stock itself is not a product. Fool me once.

The only good thing to say about the dot-com buffoonery was that it redirected bags of locked-up money into retail, sending Porsches roaring off lots and Armani flying off racks. Business had finally proven the trickle-down theory; best of all, it had put on that convincing demonstration at its own expense.

Then came the necessary hangover. World business required a sharp collective reminder that while market evaluation has a big subjective element, in the intermediate term something objective and substantive must exist for the market to evaluate. But necessary or not, the fallout was painful. *Fool me twice, shame on me,* muttered the banks—and invest-in-all-tech became invest-in-no-tech. Even the best ideas didn't matter. Capital, out of pocket through its own inattention, blamed technology rather than its own greed-binge. It licked singed fingertips and stuffed what cash it had left more deeply into the vault.

Since a pendulum's nature is to keep swinging, capital access for advanced technology has recently begun to ease. But as this happens, it's instructive to look back and see what sort of high-tech businesses *did* find financing in the lean years, mid-2000 through mid-2002. Precisely because they succeeded under such adverse conditions, the projects that found funding are arguably the soundest, best-conceived examples of newly commercialized technology in the world. Two of the clear winners in this tough race are nanoscience and its spin-offs, collectively called nanotechnology.

Even in a market-driven economy like that of the USA, fledgling technologies usually begin their capitalization with public money. Funds can be assigned indirectly through university research teams, or else directly, through state and federal laboratories. For the year ending September 30, 2001, the U.S. government budgeted $422 million for nanoscience R&D,

150 percent of the amount spent for FY 1999–2001. The Bush administration has downscaled or eliminated many programs begun by its predecessor, even (perhaps especially) in R&D. But while budgets for most U.S.
agencies supporting basic research have been frozen or pared down, the
National Nanotechnology Initiative, announced in 2000 by President
Clinton, has been continued by Mr. Bush at higher levels of funding than
his predecessor originally proposed. Allocations for FY 2001–2002 show
a further increase, to more than half a billion dollars.

Other statistics confirm the rise of nanoscience. In 1997, aggregate
U.S. spending for both government and private-sector nanoscience was
approximately $400 million. In 2001 it was three times that level; in
2002, the aggregate gain approached 450 percent. Over the period
1999–2001, the number of large interdisciplinary U.S. research groups in
nanoscience tripled to thirty. "Nano mania flourishes everywhere," pro-
claimed the journal *Scientific American* in late 2001.

The picture outside the United States is much the same. France,
Germany, and the United Kingdom have established national programs in
nanoscience. Korea, Taiwan, and China have all announced their intention to fund nano-institutes. Even Canada, whose population of 30 million equals that of California, is fast-tracking a National Nanotechnology
Institute. Administered by Canada's National Research Council and located
at the University of Edmonton, it opened provisional quarters in July
2002 and will occupy a state-of-the-art, 30,000-square-foot facility by the
end of 2003.

Pay close attention to those facts. Of all nations active in nanotech, the
Canadian initiative could pack the biggest impact for the USA. Under the
North American Free Trade Agreement, Canada and its powerful southern neighbor have established a seamlessly integrated high-tech economy.
Disputes arise in old-economy goods like wood and steel, but rarely in
advanced technology. Every BlackBerry pager comes from Canada. So
does half the transmission hardware in many North American telephone
companies. There also exists the possibility of international cooperation,
which would allow Canada and other smaller nations to pool facilities,
personnel, and IP in virtual mega-institutes.

When I flew to Ottawa to interview Dr. Peter Hackett, VP of research
for the National Research Council of Canada, he introduced me to a visitor with whom he had just concluded a three-hour meeting. It was the
president of the National Nanotechnology Institute of Taiwan.

DIE KUNST VOLLE STADT

Donald Sprung of McMaster University in Ontario finances his theoretical research and his math-based technology development by striking alliances with Swedish and Italian colleagues. As a team, they apply for basic-research grants to the European Union. In North America, the approach is sometimes called "getting money out of Brussels"—that city being the EU's administrative capital. (The international diplomatic community has a joke about poor lowbrow Brussels. Two people meet at a party. "I work for the Brussels Culture Ministry," says one. *"Enchanté,"* replies the other. "I am an admiral in the Swiss Navy.")

The more I write about science, the more I realize it's like a wedding: Funding the thing is ten times as complex as the thing itself. Nowhere do Sprung and his European associates more truly demonstrate their intelligence than in obtaining EU money for their investigations. In doing so, they have mastered something more demanding than n-dimensional string theory: decoding the mind, soul, and bias of the Brussels *uber-Eurokrat*.

"European funders aren't like North American agencies and institutes," Sprung told me. "Over on the west side of the Atlantic, institutes and agencies either buy your proposal or they else reject it. The Europeans aren't afraid to burrow into your proposal and change it to meet their specs. 'We'll fund you if you make modifications,' they say, and then spell things out in detail. They can do that because they have so many top-notch scientists on their evaluation teams."

My ears perk up; I recognize this approach. That's what U.S. venture capitalists do—*We'll give you money, but on terms that protect our investment.* He who pays the piper, calls the tune! Apparently Europe is as serious about basic research as North America is about downstream development.

I corroborate this six weeks later when I encounter a very impressive Swiss technical delegation. They hail from various places—Basel and Zurich in the north, St.Gallen in the east, Lausanne and Geneva in the south. Although they have come together from all over Switzerland, they function as a single unit, almost like a military phalanx. They represent more than their nation's separate cantons, or even its considerable investment in nanotechnology; they speak for the nation itself.

This is important. All Switzerland lives by its wits, and has for centuries. Back when I started writing about science, I heard an engineer sum up this feisty, indomitable, brainy, elitist nation this way: "Montana iron ore sells for a few dollars a ton. Smelt it into an ingot of low-carbon steel, and it's worth ten cents a pound. Process it further into chrome-moly stainless and it's worth four bucks fifty per pound. Process it into parts for top-end watches, and it's worth two hundred thousand dollars a pound."

That summarizes what the Swiss do. The watch-part story still applies, although the gears-and-wheels industry took a massive hit when digital watches based on microchips came along. The Swiss rebounded with a one-two punch. First, they developed the Swatch, a line of stylish low-cost wristwatches designed and marketed for youngsters. Second, they fashioned ultra-high-end timepieces, the Ferraris and Lamborghinis of the timepiece world. In essence, this is wearable jewelry that conspicu-ously displays success. But watches are only the start. The Swiss have fleshed out that industry with some of Europe's greatest concentrations of wealth based on high-tech. This includes biotechnology, chemicals, bank-ing, and pharmaceuticals.

Adding value in this way has been Switzerland's shtick since its con-stituent cantons amalgamated in 1521. Not coincidentally, the country has a good claim to being the world's oldest continuously functioning democracy. Self-government has long had high status among the Swiss. They've put together a federal state, which both its constitution and its citizens consider to be a free and voluntary association among sovereign entities. Switzerland may be the closest thing there is to that ancient oxy-moron, a society of anarchists. But for them it works, in every possible way. Hundreds of years ago the Swiss learned what the Iron Curtain redis-covered at its collapse. Commercial success rests on intellectual free-dom—and that in turn rests on political freedom. No vote, no thought; no thought, no money.

The Swiss have three official languages: a *Munchener hochdeutsch*, a soft and liquid French, and a bright Italian. As for their national charac-ter, it's as an exasperated Yahweh described the Israelites to Moses: "I have seen this people, and behold, it is a stiff-necked people." The Swiss are indeed unused to bowing heads and bending knees. They hate to show allegiance to any king, prince, dictator, or army—anyone they themselves have not elected.

Surprisingly, this proud and cantankerous little nation has not suffered any serious separatist movements in a long time. A bit of reflection shows why. A stone left in the open is subject to many forces that abrade it: wind, ice, water, heat, and cold. In time it falls apart. But bury the same stone deep in the earth, and it will cohere forever. The Swiss keep to themselves like a rock at the heart of a mountain. Like Israel, they are kept together by implosive forces. The Swiss have no sworn blood-enemies and are not in hourly fear of attack or subversion, but you wouldn't know that from their behavior. Switzerland deals freely with outside nations, but at the same time seems deeply suspicious of them. It cherishes its isolation and carefully chooses which foreign influences it will import.

It follows directly that the Swiss, like the Israelis, have armed themselves to the teeth. Both nations have a sting and know how to use it. Neither trusts the outside world's goodwill to guarantee its hard-won freedom. All able-bodied Swiss must serve a term in the army. Afterward they must remain on the reserve list, ready for instant call-up in the event of invasion. And woe betide the aggressor that invades. With an area under 16,000 square miles and with a population under seven million, both about the size of Massachusetts, Switzerland would seem a sad, throwaway little place. But it's nowhere near that modest, and it's certainly no patsy for a would-be aggressor. Even Hitler and Mussolini left it alone. Switzerland has earned its national symbol of a rampant bear.

The Swiss have spent decades excavating enormous fortresses deep inside their mountains. All major roads have built-in tank barriers. Normally these are lowered flush with the road surface, but they can be remotely raised from central command posts at an instant's notice. Armored columns so immobilized would be sitting ducks for airstrikes. In addition, hardened artillery emplacements hidden in the mountains already have the coordinates of the tank traps punched in to their fire-control computers. An invader's tanks would be scrap metal in half an hour.

Switzerland is unlike Israel in one thing: It has no territorial ambitions. It's content with the geography it has. Yet this independence does not extend, as it does in so many of the world's smaller nations, to isolationism. Switzerland is a textbook case for advanced economic geography. The Swiss maintain their high per capita standard of living—about 50,000 Swiss francs per year for every man, woman, and child in the city of Basel, for example—by making and exporting high-priced items. In adopting new ideas from beyond their borders, the Swiss lead the world. They locate hot new technologies, sweat blood to become proficient in them, and then ramp them up to be massive moneymakers. Throughout this process the Swiss are high-graders. They import only the best. The whole culture is a perfect Petri dish for nurturing new ideas like nanotechnology and its kissing cousin, MEMS, or micro-electro- and mechanical systems.

Basel is a small city of about 200,000 located in the exact geographical and commercial center of Western Europe. It houses the headquarters of CIBA, Novartis, Roche, Syngenta, Clariant, and the Lonza Group. Each of these huge multinational pharmaceutical firms has aggregate yearly sales in the tens of billions of U.S. dollars. One or two are nearing the hundred-billion mark—bigger than IBM.

Basel is neat, clean, prosperous, and tidy to an obsessive degree. They've stopped scrubbing their front steps in Amsterdam, but by God, they still do it in Basel. The city straddles the River Rhine not far from its source in the Swiss Alps and, I kid you not, even its *waterfront* is tidy.

There are those who denigrate this. The most famous sneer came from Orson Welles in the classic film *The Third Man*. Switzerland has had four hundred years completely free of want, pestilence, or warfare, Harry (Welles's character) tells the hero. It's been prosperous and stable all that time. There's never been any real threat to the place: no starvation, no invasion. And what's the greatest invention they ever came up with? The cuckoo clock. Fine, says Basel, mock away. Who's smart? Who's peaceful? Who's rich?

Dr. Alex Dommann has a demeanor that's perfectly Swiss. Erudite, multilingual, intelligent, focused—and unsmiling: almost grave. Not much personality visible on this sleeve. In this, Dr. D seems a clone of his fellows in a Swiss trade delegation to North America, who are here to talk about nanotech. When Dr. D gets intense about a technical point, his eyeglasses—as the Canadian humorist Stephen Leacock said about his Greek professor—glitter with excitement. But his first words show why the Swiss are becoming a powerhouse in nanotech. It's all to do with specialization. Already they've done a study and sussed out the rest of the world. That's shown them where they can most profitably fit in.

"Japan's main nanotechnological interests are in nanodevices and consolidated materials," Dr. D announces, as if addressing a room. "Europe and the USA lead in bio-nano, as well as in nanotech-based materials science. The United States occupies sole number-one position in nanosynthesis and assembly. Also in materials with high surface area." And the Swiss? "We are considering a concentration in an area where we have already achieved excellence. That is fluidic microchannels and nanopore filters." He pauses. "Fluidic *nanochannels* and *angstrom*-pore filters, I should say. "

Dr. D's self-correction makes his utterance more accurate by an order of magnitude: 1 nanometer = 10 angstroms. A tiny smile betrays his delight in such precision. Fine, very impressive; but how might such technologies be used? "We are already investigating a commercial application," Dr. D answers. "We think we can shortly market a home pregnancy test that gives results in only ten seconds. The woman's urine flows into a myriad of molded nanochannels, and the test reaction takes place in these. The reaction detects a pregnancy protein." Aha! I say. The detected protein exists in tiny amounts, right? And using this nano-application means that even those tiny amounts give reliable, unambiguous results? I'm rewarded with another smile, as if the professor were patting an earnest young student on the back. "You have identified an important advantage, yes. The test works very well with only the quantities of protein found naturally in the body."

I make a note and underline it. This new Swiss nano-app could be one of those solutions that, like the lever or the calculus, transform fields far beyond the immediate problem for which they are devised. Biotechnologists have been wringing their hands about the difficulty of detecting femtograms of natural protein. Nanochannel assays could be the way. A technique called PCR lets them multiply a few atoms of DNA into much greater quantities, which can then be tested. No similar technique has yet been found to multiply proteins. But the Swiss nano-application may mean the long-sought "protein PCR" is unnecessary. If a test is sufficiently sensitive, there's no need to increase the quantity of what it detects.

While the Swiss may have brought the manufacture of molded nanofluidics to a high pitch, they aren't the only players in this field. "Soft nanotech," as the science of molding ultra-smooth surfaces has been called, has a number of Americans at or near its top. Among them are Dr. George Whitesides of Harvard and Caltech's Dr. Stephen Quake. Whitesides is the elder statesman of soft nano. At only thirty-four, Quake is its bold young visionary, and a successful entrepreneur to boot. Quake started working with nano-molded surfaces while still a student at Stanford. His spin-off firms, Mycometrix and Fluidigm (pronounced "fluid-dime"), have progressed beyond the one-test technology of the Swiss pregnancy test. Quake et al. are now on the trail of more complex devices. These may do for nanofluidics what Intel did for microelectronics: concentrate an entire laboratory on a single intricate chip. (See Chapter 7.)

Fluidics is the science and technology of moving fluids through progressively smaller conduits. This manipulation is done for various reasons. In applications such as the Swiss test, the fluidic device examines the fluid. When Dr. Quake's devices are involved, the fluid may be a means of computation, like electrons in a microchip.

What the Swiss may lack in cutting-edge theory, they make up for in sheer accuracy. And in nanotechnology, more than in any other human endeavor, God is in such details. Precision of manufacture may make the difference between an application's failure and its market success. The Swiss have identified a subarea of molded nanofluidics that not only penetrates but dominates its market. For Switzerland, soft nano could prove to be the third millennium's commercial equivalent of the wristwatch.

Interestingly, too, soft nano may be silicon's more enduring contribution to nanotech. Semiconductor materials based on doped silicon will probably be obsolete in ten years. (Doping involves modifying the silicon's properties by adding trace amounts of other substances such as germanium.) But silicon also underlies the flexible, thermoplastic compounds that accurately take on nanoscale features. Silicon can be molded into nanoscale devices almost as readily as nylon or polyurethane can be molded into toys and toothbrush handles. In these molded nanodevices, it is silicon's striking ability to copy its mold that matters. By and large, emerging nanotech has little use for silicon as a semiconductor. This application of silicon may persist when the silicon transistor is only a memory.

"Then there are the bulk metallic glasses," Dr. D continues. "These were first found at Caltech." (The U.S. connection again!) "They are most interesting materials. They exhibit high elastic strain and high tensile strength, like commonplace metals. And they have excellent corrosion resistance. There is only one mechanism of mechanical failure due to strain." Uh? "I mean, there are no internal dislocations in the microstructure or the nanostructure of these materials." Thank you. "Also, the bulk metallic glasses appear to be amorphous, that is to say, noncrystalline. Their surfaces are extremely smooth. Even under high magnification, they look like liquid mercury at STP." You mean the oil additive? "I mean *standard temperature and pressure*," Dr. D says, looking severe. "These bulk metallic glasses make excellent molds for producing our nanofluidic channels. In addition, there is a blood-cell-counting device under development."

Then, as I take notes, I get the giggles. It's appalling. I feel like a kid in a library, shushed by a lady who wears sensible shoes and has her hair in a bun. The more I choke it down, the more it beats up on me. *In-te-ress-tingg. Quan-ti-tiesss of prrr-ohh-teee-innnnn.* In this learned gentleman's mind, *blood-cell-counting device* may be a single word, *Weissblutzelwissen-schaftennumerator* or something . . . . Damn, this is juvenile! Dr. D's English is light years better than my German. But just as I get myself in hand, my brain betrays me with six seconds of mental video from *The Simpsons* episode where the Germans buy the nuclear plant. *I haff been selectedd to deal vith you because I vill remindt you of ze lovable Sergeant Schultz of* Hogan's Heroes.

An old memory comes to my aid. Years ago I was lofted ninety feet into the air on an ancient cherrypicker with a rickety safety barrier. Some astronomers were showing me the giant steerable parabolic antenna that eavesdropped on the snap and hiss of distant galaxies. My heart was hammering with terror as I rose in the air. Then I looked into the viewfinder of my camera to take a picture—and everything was fine. My organs filed back into place in orderly fashion, and I could do my job. A similar thing now happens here. Concentrating on what's being said rather than how Dr. D says it, I lose my giggles. Something he's just mentioned catches my interest; I come back to it.

"You said there were other possible uses for devices with nanoscale features obtained by molding?"

The spectacles glitter. "Ah! Yes. Certain molded forms may function as parallel waveguides for electromagnetic radiation in the visible spectrum."

"Why do you want a visible-light waveguide? That's LOS, right?"

"Line-of-sight transmission, correct. The European Space Agency is investigating such frequencies for independent communication among orbiting satellites. Such is not now possible. Radio signals travel *to* the earth *from* one satellite, then up *from* earth *to* another satellite. With our nano-manufactured waveguides, two or more satellites could communicate directly. The waveguides use the shorter wavelengths, which permit greater information densities than RF. Thus the satellites may exchange more data, in shorter time, directly." To understand how big a breakthrough has been brought about by the Swiss' precision nanotech, imagine what naval logistics would be like if ships could communicate with one another only via radio operators on shore. Modern trade and military strategy would be crippled. Direct intraspace communication promises a similar advance.

The United States and Europe, Dr. D tells me, have different approaches in turning basic nanoscience into commercial technology. The U.S., he says, spins off large numbers of small new start-ups whose target customers tend to be the big, established firms. The Europeans try to influence their big firms directly into exploring and applying nanotech themselves. To date, the Swiss have enlisted the electronics giant Siemens and the big pharmaceutical firm Naxus in their program to convert experimental nanoscience into marketed nanotechnology.

All this paints a curious portrait of two national characters: Yankee go-getting entrepreneurship vs. the sober, organized, collective march of Western Europe. The Yanks make the Euros look stodgy. The Euros make the Yanks look haphazard, disorganized, and all over the map. The market will sort it out. Or else it won't; both approaches have their place.

I get a better sense of this when Dr. Hannes Bleuler, another member of the trade delegation, briefs me on a Swiss federative program in nanotech. This project seems more formal than the North Americans' networking style, which is less planned and more ad hoc. Dr. Bleuler (pronounced "*bloy*-lur") is from EPFL, the Ecole Polytechnique Fédérale de Lausanne. The project unites many smaller projects that have already sprung up in various Swiss labs. It will investigate a concept that isn't yet close to reality: the nanofactory.

Consider the average factory of today, Dr. Bleuler tells me. "Ninety-nine percent of the space and energy it consumes is waste. The energy goes to nonmanufacturing uses, such as heating and cooling vast areas for human workers and for machines that are larger yet. Why should we do this when the output we desire is small? One could imagine a factory to produce, say, nanoscale transistors, a whole factory with power and materials supply, that is one-tenth of a cubic meter in size, no larger." Now *there's* a neat idea: a nanoshop no bigger than a breadbox.

This is still only a vision, Dr. Bleuler cautions me, a great and difficult goal. "We are working toward it from the top down. That is to say, we start at the mesoscale and steadily decrease the average size of the operation— how big its components are, how big are the components that it makes." Any progress to date? Dr. Bleuler's reply nearly flattens me: "We have produced a functioning atomic force microscope that is one inch square. It is laser-cut from a sheet alloy substrate and is a thousand times less expensive than today's AFM. This sensor has already attained atomic resolution." Well, shut my mouth! Serves me right for mocking these guys.

Federative projects such as these have arisen because the pragmatic Swiss think they offer the best chance of continued prosperity in an uncertain future. Individual labs may have some good ideas, but a network of labs is more likely to see each concept through to the marketplace and earning money. That's important to the commercial excellence that underpins Switzerland's long-established autonomy, which has allowed them, thus far, to avoid membership in the European Union (EU).

Oddly enough, that very autonomy may soon come under pressure from nanotech. Federation and integration are occurring not only within the various Swiss laboratories; they are increasingly bypassing all national borders. Swiss nano-projects increasingly tie this scrappy little nation more closely to the EU. Bilateral agreements signed in 2002 already permit the free flow of funds and labor among Switzerland and the nations of the EU. In mid-2002, the EU supplied one Swiss worker in six. It's a paradox, but it seems inevitable. To stay independent, the nation must decrease its standoffishness and open itself up to more technical and economic alliances.

By 2015 or so, in fact, dissolving borders and blending nations may prove the greatest social achievement of the nanocosm. Science has long been international; nanotech might be the strongest leveler of all. If so, it would recapitulate what it is already doing among the various older disciplines in academia. The study and sale of atoms may yet remanufacture the earth into one hegemony: a single prosperous state. For the first time since empires arose five thousand years ago, our planet could become the serene and peaceful entity that it appears to be from space.

THE HONEST-TO-GOD NANOMACHINE

Remember the nanomanipulator—a nanoscale robot that intervenes at the atomic level and constructs complex devices atom by atom? One of the craziest things about the nanoboosters' concept of a nanomanipulator is that it's unnecessary. We don't have to wait decades or centuries to invent one; it exists today. It's not in the form that the boosters imagine, and probably never will be. But in performing nearly every function that a nanobooster's quirky mind can conjure, the nanomanipulator—also known as the molecular assembler—is already here. It's our old friend, helper, and antagonist: the hero that constitutes our bodies and the villain that tears them down. It's food and eater, enemy and friend; as new as tomorrow and as ancient as the stars. It's the molecule.

Science defines a molecule in less poetic terms. To those in the laboratory, it's a neutrally charged aggregate of two or more atoms that under certain circumstances continues in a constant configuration over time.

Nature exists in layers; and to some scientist or another, every layer that exists is fascinating—from the Planck radius of 10^{-33} cm, to the line connecting us to a quasar, or 10^{28} cm. (As a throwaway fact, the ratio of these extremes is 1:1000000000000000000000000000000000000000000000000 0000000000000.)

With this in mind, I go to talk with Dr. Neil Branda, a brand-new associate professor of chemistry at Simon Fraser University in Burnaby, B.C., and the proud possessor of a Ph.D. from the Massachusetts Institute of Technology. Branda is in love with molecules, and among the cognoscenti he is considered something of a hotshot. Behold the signs of the hotshot, that ye may know him: A press release from the school that's just nabbed him, extolling his virtues as "a rising star in organic materials science." An equipment grant equal to roughly ten times his first year's salary, to help him set up his new lab. A brigade of loyal graduate students who have followed Branda from his old home to his new one. A rigorous experimental program, with strict methods but vainglorious aims. Wild ideas about molecules, and equally wild results.

"First off," Branda says, striding into his corner office and speaking to me even before he sits down, "you have to understand where this group is coming from. Our sole interest is governing the function and behavior of single molecules. We're a bunch of control freaks."

Branda and his team, he tells me, approach individual molecules as components of nanomachines. They start with the Middle Kingdom, the macroworld, and look at how full-sized machines work.

In the nanocosm, as in all of nature, structure and function are intimately connected. Thus Branda and his team are especially interested in controlling the links between how things are shaped and what they do. "If you can control the structure of molecules," he tells me, "you can also control their functions. We're starting with molecular structure; we'll get to the specific applications later."

Unlike the nanoboosters, Branda's team is not out to reproduce belt sanders and escalators at the atomic level. Branda regards that approach, the Drexlerian approach, as crude and unimaginative, a source of quiet amusement. "Look, Daddy, I made a time machine out of tin cans and Plasticine!" "That's great, honey . . ." Branda *et al.* take a more sophisticated tack. They approach nature as, well, scientists. After they examine the functions of large-scale machines, they design individual molecules that can duplicate or improve on those functions at the nanoscale: for example, catalysis (i.e., making things happen) or negative catalysis (i.e., instantly turning chemical reactions off in midflow). Modification, shaping, milling, moving—every one of these processes is on the Branda team's to-do list. One by one, they're being achieved.

"We're academics," Branda tells me. "We go at things like academics. We're interested in what's happening at a very basic level." That being said, Branda admits he can never rest content with discovering. He has to tweak, change, modify, and meddle. I hear this a lot from chemists. Something in chemistry tempers some of the scientist's awe before the face of nature, substituting a little-kid urge to toss in sticks and see what happens.

Perhaps because of this practical, experimental attitude, Branda and his team invariably always keep one eye cocked for possible commercial applications. He gives a raft of examples; one sticks in my mind. A big chemical company produces many industrial materials by catalytic reaction. In most cases the reaction is tightly controlled, but when there's a glitch the results can be horrendous. If catalysis goes awry and a product suddenly changes from the desirable cascade of small beads to a single lump, a chemical reactor costing tens of millions of dollars can instantly fill up with solid polymeric gunk. There have been instances when the only way to clear a system that gets constipated in this manner is to go into it pipe by pipe with drills and chisels—not an elegant solution, to be sure. But Branda and his kids are on the trail of "switchable molecules" that could halt so drastic an outcome the second it shows signs of happening. The intervening molecules would switch off catalysis and give the production engineers time to pull their system back from the brink of disaster. The result: no more factories filled with useless gunk when reactions go awry.

"Any chemist can make A react with B," Branda says. "But what if you don't want 'em to? It would be nice to have a chest full of molecules you can turn on and off at will."

Another possible application of Branda's approach is drug delivery. Branda sees no need for enormous molecules to transport and dump pharmaceuticals, whether these molecular vehicles are natural or synthetic. To him, dendrimers are interesting, but unnecessary. There are more elegant ways to FedEx a drug—and to control its chemical activity at the same time.

"How about a nice little custom-designed molecule with floppy arms like pincers?" he asks. "Something hardly bigger than the drug it's carrying, maybe not even that large. It gets to its target area, opens up like a dump truck's tailgate, drops its load, and returns for more. By whatever standards you want to apply, that's an honest-to-God nanomachine." There's no need for sleeves and bearings and rotors and God knows what macroscale-mimicking mechanisms, all looking like their counterparts at the macroscale. All that science-fiction stuff imports its design from the macroworld, reduces it a few million times, and then has to be hard-engineered from diamond. Why reinvent the wheel? Especially when wheels may not even function at the nanoscale. Nature's already supplied us with molecules; let's use 'em.

Branda puts his sandals on a window ledge. His legs alone are longer than I am. "There's no need for all that Drexlerian stuff," he says. "We're already doing what they're only imagining. They want to incorporate atoms and molecules into nanodevices. Our molecules *are* nanodevices."

Doing this type of work effectively, Branda says, means thinking outside the box of strict analytical or synthetic chemistry. "You have to step back from the atoms from time to time." For instance? "Protein function depends to a large extent on shape. So some groups are looking at the mathematics of knot-tying and topology," he says. "There may be ways to improve on nature here. We want to understand what nature does, but we're really after more elegant ways of making things—products, shapes, proteins—to order." More elegant than what? "Than nature, of course."

Branda exhibits a chemist's approach to biomimicry. Sure, let's start by understanding nature. Let's duplicate natural nanomachines and processes in the lab. But why stop there? Let's go one better on life.

"Biological systems are the best examples of nanotechnology out there," Branda says. "In one sense we'll never improve on them. But in another sense, those natural systems are often costly in terms of energy. They haven't been designed; they've 'just growed.' Your average enzyme works fine as a natural catalyst. But it's usually an enormous molecule with a little-bitty active site that does all the real work. Surely we can optimize that."

Maybe enzymes are already optimized, I suggest. The active site on a pair of pliers, its grip area, is small; but big handles are needed for leverage.

Just as there's no such thing as a pliers that's all grip area, there could be sound chemical and energetic reasons for all that enzymatic bulk.

Branda shrugs, unimpressed. "I still think we could improve a lot of natural enzymes. It's worth a try."

Yet another of Branda's interests (as he describes himself as "interested in everything and easily distracted") is artificial photosynthesis. Two words, nine syllables: simple concept, tough to achieve. Branda wants to do what plants do, on demand, using synthetic molecules. That would provide an everlasting power source for his molecular nanomachinery. Doing this won't be easy.

"Natural photosynthesis is incredibly complex," Branda admits. "It has proven very difficult to duplicate using biomimicry." Again, his interest lies in regulation: How do you switch photosynthesis on and off at will?

Branda's team is also investigating a class of molecules called porphyrins, which are distantly related to chlorophyll. Like that molecule, porphyrins capture and store light energy. "They are energy funnels," Branda tells me. "They grab light energy and dump it elsewhere."

Branda et al. use other molecules called photochromes to signal porphyrins' successful light-harvest. These signal molecules change color when they receive light energy, giving the scientists a clear indication of success.

Since energy storage and consequent color change are both reversible, the Branda group's synthetic molecules might also support ultra-high-density memory. This wouldn't be set-and-forget, either, like IBM's experimental Millipede ROM device. The Millipede, announced in 2002, is a variant of those governmental check-signing machines that let officials autograph a hundred documents at once. Like those machines, the Millipede looks like a claw. It has dozens of arms that simultaneously gouge nanometer-long tracks in a smooth surface. In place of this permanent record, Branda's porphyrins offer the possibility of real, nanoscale, erasable, rewriteable, molecular RAM.

Why stress applications? I ask. Branda shrugs again. "Maybe I'm not that traditionally academic," he says. "I get sick of writing in journal papers 'it would appear this effect may have the potential to do such-and-such.' If we can go those final steps and demonstrate proof of concept, we will."

Branda's group sends the porphyrin-based molecules it synthesizes all over the world, for other groups to investigate and apply. Topping his mailing list are the Scripps Institute and the University of California at

La Jolla. Other partnering laboratories are as far-flung as Phoenix, Arizona, and Bologna and Siena, Italy.

"We like to farm stuff out," Branda says. "We're best at producing these molecules, but others are best at using them. Everyone benefits from that kind of sharing." Besides, he says, these days you have to work this way. "If you go into a grant process without partners, you're fried."

HEAT

The trouble with nature is her rigor, her consistency. If only she weren't so damned *logical*. If we could break free from her insistence on sequence and consequence, we could do amazing things. Teleport matter. Exceed lightspeed. Make complex nanobots. Understand rap.

Alas, nature is relentlessly consistent. All apparent violations of her iron order prove to be illusions. Some of these surprises, the scientific equivalent of stumbling over an unseen object, prove to be portals to new realms. This is what really gets scientists' goat about not understanding something. Maybe that anomalous fact they can't explain is not an artifact but a clue: a door to a brand-new paradigm, a gateway to another dimension. You can't tell unless you chase your stray fact down, corner it, tie it up, and explain it.

One of nature's most insistent rules involves heat. Sooner or later, say Lord Kelvin's famous laws of thermodynamics, all energy ends up as heat. Electricity, gravitational potential, momentum, fissile isotopes, sunshine: However energy makes its first appearance, heat is the final result. This heat cascade is irreversible. Because of it, everything will ultimately exist at the same low temperature. Cosmologists call this the heat death of the universe. After a few hundred billion years, no stars or galaxies will shine; all matter will be as cool and dark as the space that surrounds it. In energy terms, this will be perfect communism, with every particle the equal of its neighbor. (*This is the way the world ends,* wrote T. S. Eliot, *not with a bang but a whimper.* Eliot wrote his prescient lines long before cosmologists predicted universal heat death. Poetry, the poor man's physics.)

It's not all *Götterdämmerung.* Just as death gives meaning to life, heat death gives the universe its overall direction. Clocks run down, but only after they've been running a while. Same with car engines, radios, computers, and stars: Unless heat accumulated, nothing would work. Never mind we're moving toward the graveyard. At least we're moving.

Heat is the random motion of invisible particles—from molecules and atoms to subatomic flecks like protons. As such, heat is often portrayed as the prototype of chaos. Scientists even measure disorder by the amount of heat that a given mass contains, an expression known as entropy.

Of course there's a paradox in all this. (Nature loves paradox more than anything; it's the surest way to drive a scientist nuts.) The conundrum here is that a certain amount of disorder is necessary to reorganize any existing reality into new order. You see this law everywhere: Inflation ends stagnation. You can't make an omelet without breaking eggs. Dynamite makes the railway. (Although I'm not sure I agree with a bumper sticker I recently saw in Oregon: THERE IS NO PROBLEM THAT CAN'T BE SOLVED WITH HIGH EXPLOSIVES.)

The Principle of Necessary Disorder extends beyond mechanics into politics. Freeze expression, and you freeze out new ideas; freedom and chaos go hand in hand. This idea shattered the Soviet Union and has most big corporations (and the People's Republic of China) in its sights.

Heat can be dangerous: too much destroys living systems and can even shatter atoms. But too *little* heat, and matter stays forever dead. At absolute zero atoms never shake, rattle, and roll enough to detach themselves from their crystalline prisons and get to work combining into interesting chemicals. It's like a party—one neurotic works wonders, but invite one too many and all hell breaks loose. Where heat's concerned, the trick is knowing how far to go.

This is admittedly TMA, Too Many Analogies, but all of them bear on how global business will develop the nanocosm over the next ten years. It's fine to talk of molecular assemblers and angstrom-level electronics, but there's a major problem at these scales. As it is with living systems, so it is with those artificial systems called computers that now border on life. A little heat is necessary; too much fries brains, both natural and synthetic.

In all technology, but particularly in nanotechnology, the smaller you get the more heat becomes a problem. As Simon Haykin noted in his Purdue lecture, a modern microchip can generate ten watts of heat per square centimeter when it is operating. That's about one-third the thermal concentration of a kitchen toaster. Remembering that entropy is the quantity of heat in a given mass, ever-greater entropy (i.e., heat output per unit of mass or surface area) awaits the engineer who rappels down from macrocosm to nanocosm. While smaller components put out less *total* heat, their heat *concentrations* rise out of control. Results may include fatal malfunctions from overheating in hardware, software, and IT systems.

Strangely enough, things used to be worse. Before solid-state electronics, digital computers were vacuum-tube monsters that filled rooms. The electrons used in these old electronics were "thermionic"—that is, they were produced by heat. ENIAC and BRAINIAC required vast, dedicated air conditioners that sucked away hundreds of thousands of BTUs from every mainframe. The entropy levels of the old computers, in the sense of heat produced per unit volume, weren't that high. But their size was so great that their total heat output was enormous. It's no coincidence that engineers started wearing short-sleeved dress shirts when the big machines came into use. Sharing a room with your typical 1950s mainframe was too hot for long-sleeved workshirts. (As a sociological aside, the free fall in adult smoking levels since 1960 owes less to epidemiological studies and aggressive public-health campaigns than to magnetic media's intolerance for smoke. In that way, at least, it didn't hurt us to reify ourselves and imitate our machines. Our floppy drives knew what was good for them and us, better than we did.)

While solid-state hardware (i.e., transistors) replaced vacuum tubes and solved the immediate issue of total heat output, the greater problem of entropy—that is, more heat in less volume—did not go away. Even today, many a microchip-based computer cooks itself when operating too far above room temperature. This is only going to get worse as components get smaller. Carbon nanotubes, which many researchers consider front-runners in the race to achieve nanoscale transistors, don't just conk out or melt when they get too hot. In the presence of oxygen, they explode. *Bam!* Bye-bye nanoprocessor. Your computer won't reset when it cools down, sir; it's blown its CPU.

Nanoscience, however, is rapidly showing us how to control heat within the nanocosm. The technical and economic consequences are major and direct.

In theory you could cool a VLSI nanochip that contained ten trillion transistors and diodes simply by operating it in a blast of dry air, cooled to zero Fahrenheit. This is workable for mainframes and large fixed emplacements. It is not a practical option for portable equipment such as notebooks and BlackBerries. Researchers are therefore turning their attention to cooling devices that can be built directly into a nanoelectronics-based machine.

Possible solutions to the heat problem are emerging from experimental and theoretical nanoscience. In a 2002 paper published in *Applied Physics Letters,* Luis Rego and George Kirczenow propose adding a second current to MEMS or nanoscale electronic devices. They call this effect "classical" because it has a precedent in macroscale physics: that is, the cylinder in a diesel engine. Just as the hot, exploding gas in a diesel cylinder loses energy when it pushes against a piston, electrons moving inside MEMS circuitry could be made to lose energy by pushing against a counter-current, separate from the principal current that shuttles and encodes the information. I'm suspicious of this suggestion. As every municipal engineer knows, waste must go somewhere. Energy may be extracted from moving electrons at the nanoscale in this way, but it still must be disposed of by being dumped somewhere. In the diesel, excess heat goes out the tailpipe with exhaust gases. Where would the heat extracted by the counter-current go?

Another, likelier solution comes from a company called Cool Chips PLC. Its technology is well beyond theory and into the benchtop stage. It involves extreme refinement of production methods, like those used by IBM when it makes its hard-drive heads. This is quality control on the nanoscale.

Cool Chips' working models of thin, solid-state wafers can be inserted directly into solid-state devices to provide cooling. Eventually, suggests a company spokesman, this technology will be able to "replace nearly every existing form of cooling . . . [including even] air conditioning." You have to hand it to Cool Chips' developers; they're not shy. But maybe they don't have to be. Devices using their cooling chips tip the scale at only one-tenth the size and weight of conventional air conditioners based on compressors. In addition, thermal efficiency of the new chips is 70–80 percent. This compares with 40–50 percent for the best available compressor-based systems and only 8 percent for older solid-state cooling devices called thermoelectrics.

Because Cool Chips are modular—that is, they work individually or in gangs—the company believes they will replace thermoelectrics and compressors altogether. One tiny microchip a few millimeters square could cool a "terasistor"—a nanochip with a trillion transistors built in. A Cool Chip one-inch square would provide the cooling power of a standard refrigerator; a 5" × 5" VLSI array could air-condition the average detached, three-bedroom house.

Cool Chips are still under development, but they promise a lot. No, they won't arrest or reverse the heat death of the universe. Cooling is localized; on the whole, more heat is produced. But the guts of a notebook with a nanotech CPU will stay optimally cold, even at noon at the equator. And in five years or so, your car's air conditioner could shrink to the size of a sugar cube. In the cosmic short term, say the next two hundred billion years or so, that's good news.

Here's how the thing functions. A working Cool Chip will get very cold on one side and very hot on the other. So do thermoelectrics. But more important, the Cool Chip stays that way, with *aah!* on the obverse and *yowch!* on the reverse. By contrast, thermoelectrics tend to keep remixing hot and cold. That makes them energy-inefficient power guzzlers.

When I review the technical papers on thermoelectric function, I'm reminded of de-gnoming the Weasleys' garden in J. K. Rowling's *Harry Potter and the Chamber of Secrets*. It's an endless process, an exercise in frustration. As fast as you chuck the gnomes out one door, they sneak back through another one. Changing the garden's Gnome Equilibrium (G_e) to a lower level requires constant vigilance and relentless work.

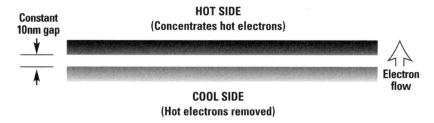

Figure 3-1. Quantum tunneling in the Cool Chip

Cool Chips slams the door on the evicted heat gnomes, so they stay out for good—"gnomes" here being hot electrons. The firm's main innovation is in the barrier that separates the thermoelectric wafer's hot and cold sides. It's not a thermal or an electric insulator. Those are material substances, usually packaged air or inert gas such as neon. The Cool Chip gap is literally immaterial. It's not a solid, liquid, gas, or plasma. It's a gap, a vacuum, *nada*. Electrons cross the gap in one direction only, using an odd effect called quantum tunneling. They're assisted in their escape by a small electric current that steers the energetic electrons, and the heat they hold, to the chip's hot side (see Figure 3-1). The same tiny current also keeps the hot electrons from tunneling back. Because there's nothing in the insulating gap to conduct them, they can't sneak back under the hedge like Rowling's gnomes. The hot electrons stay put where they belong, on the chip's hot side, and so does the heat they carry. In standard thermoelectric systems, by contrast, the high-energy electrons quickly return across the gap via conduction. Almost as fast as the system separates heat and cold, the two remix.

Cool Chips' special technology achieves in practice what was long thought to be an impossible abstraction, a thermodynamic thought experiment. It creates two surfaces that extend over several square centimeters, are separated by a constant gap of 10 nm, and never touch at any single point. To give the same achievement in macroscale terms, that's like constructing a huge room 10,000 km square (6,250 miles), with no columns and with a perfectly flat floor and a perfectly flat ceiling a constant 10 feet above it. In case you were wondering, that's nearly the diameter of the earth. And except along its perimeter, the room would not have a single pillar or support over all its twenty billion acres of floor space.

Even as prototypes, Cool Chips' wafers have a lot of room to accommodate future innovations in computer nanotechnology. A gap of 5 nm, for example, calculates out to carry a theoretical heat-extraction limit of 5,000 watts per square centimeter. The next generation of mesoscale devices will probably produce heat at only two percent of this level. Current microchips' heat output is ten times less again. Cool Chips may have come along, like the computer itself, at exactly the right time. Freed from the effects of too much localized heat, nanotechnology can progress that much more rapidly. China, Japan, Korea, even the U.S. Sun Belt—all would still be undeveloped without air conditioning. The nanocosm is much the same.

As I check figures and punch my calculator, I find myself wondering if Cool Chips has not come close to another thought experiment— arguably the most famous ever made. It's the brainchild of the great Scottish physicist James Clerk Maxwell. About a hundred years ago, Maxwell imagined two adjacent chambers, each containing gas and both sealed from the outside world. The chambers are also sealed from one another, except for a single gate the size of a gas molecule. Guarding this gate would be an intelligence with one power: the ability to discriminate between fast and slow molecules. It would pass slow molecules in one direction only, fast molecules in the other direction only. The net result would, after a time, be one chamber that contained only slow-moving molecules, next to a chamber that contained only fast-moving molecules. In other words, the chambers would no longer be at the same temperature: One would be cold and one would be hot. Should such a "Maxwell demon" ever prove possible, its discriminatory intelligence would reverse the apparently irreversible flow of entropy, which inevitably has everything existing at the same low temperature. Thanks to Cool Chips, perhaps the heat death of the universe is not so inevitable at all.

Cool Chips' proprietary systems are completely silent and use no motors or other moving parts. "Once heat is trapped on one side [of the chip]", says company CEO Isaiah Cox, " it cannot easily return." The hot side radiates its energy to a heat sink, cooling the other side and the nanoelectronics it protects.

Why hasn't such quantum thermo-tunneling been done before? "Nobody imagined it was possible to get large surfaces areas close to each other without making occasional contact," Cox tells me. "Our scientists imagined a way to do it, then they accomplished this goal." Test machines have been completed, and production design is continuing into 2003–2004. The company's niche is cooling, but its technology seems red hot.

THICKER AIR

Dull stuff, air. Mostly molecular nitrogen, which is all but inert until lightning catalyzes it, and the oxygen it's mixed with, into nitrates suitable for fertilizer. Twenty percent oxygen, some carbon dioxide, a few trace elements. Bo-ring.

But while air may not sound like much, it has one surprising trait: It's a slippery escape artist. It can infiltrate through practically anything. Leave it a gap only a few nanometers wide, and it will vanish like a thief in the night. That has enormous consequences for many industries and most people, everywhere around the world. Wonder why you need to service your car only twice a year, yet need to check tire pressure weekly? Why those three small tennis balls in that $1.99 container you got last month loaf across the net, despite your supersonic serve? Why that $5 helium birthday balloon needs Viagra after two days?

It's air leakage, friend. Not *leakage* exactly: that implies a visible hole or an audible hiss. This is exfiltration—the tendency of the molecules in air to find and exploit every escape route that's as large as they are. Pressure increases exfiltration, because it forces out molecules that much faster.

But (you say) aren't there air-proof barriers? Yes, but they tend to be heavy, crystalline solids such as aluminum, steel, and other metals. At the nanoscale—which is what counts for air under pressure—compounds like latex (for balloons) and vulcanized natural rubber (for tennis balls and car tires) offer countless escape avenues for wayward air molecules. They are less secure than a federal prison; they're more porous than a piece of Swiss cheese.

Enter a new example of commercial nanotechnology: a workable way to keep air penned up, for more hours, at higher pressure, than ever before. It's a product of a new firm, InMat. Nanotech tends to think in years, nanoscience in decades; but this company shows the technology how to survive and prosper in the short term, quarter by quarter. In this way it can earn the dollars, and more important, the time, to stay alive until its greater miracles can occur.

InMat is the brainchild of Dr. Harris Goldberg, an applied physicist from Hillsborough, New Jersey. A compound his company developed and tested, marketed under the trade name Air-D-Fence, was recently chosen by sports giant Wilson to help seal its top-of-the-line Dual Core tennis balls. It didn't hurt InMat that this product was chosen as the official ball for the 2002 Davis Cup, a global yearlong tennis extravaganza using thousands of the new balls. Having become a trusted original equipment manufacturer (OEM) to a big firm with high profile, InMat moved closer to gaining toeholds in other air-retention applications. In time, these other applications (car tires, for example) may prove far bigger than sports balls.

InMat's expertise lies in custom designing and then producing ultra-thin coatings that retard air exfiltration by orders of magnitude over standard rubber compounds—even thick, expensive butyl rubber. The Air-D-Fence coating that lines a Wilson Double Core ball is only 20 microns thick, less than a thousandth of an inch. The weight it adds to the ball is negligible: below the ball's manufacturing tolerance. Yet so efficient is this coating at retaining air that Wilson guarantees its balls will keep their factory-inserted overpressures of 13–15 psi for at least four weeks, which is twice the expected lifespan of any other tennis ball. Wilson has also beefed up its new ball's exterior fuzz. This and the

Air-D-Fence combine to create a ball that lasts through a set of Andre Agassi serves, or many weeks of mom-and-dad games at the local park.

Goldberg began InMat in 1999 by imitating Moses and leading his entire R&D group from DuPont *en masse*. He describes his people as "having incredible loyalty to InMat . . . There's no employee turnover to speak of." Each staff member is expected to take on multiple roles, filling in for colleagues as needed and helping out wherever necessary. Probably because it was not so much a start-up as a continuation, the company took only 28 months to develop three new market-ready products.

"Our product-cycle time," Goldberg states laconically, "is very brief." How brief? InMat, says Goldberg, routinely evaluates proposed changes to its coating properties in a few days.

Goldberg's application has been called an example of "passive nano-technology." Air-D-Fence doesn't have the complex goals of active nanomachinery or molecular manipulation. Instead, it solves a long-standing problem, air exfiltration, by placing a tough nanocomposite barrier in its path.

Air-D-Fence contains particles of vermiculite, a white material that in much larger chunks is sometimes put into potting soil to increase soil aeration. Goldberg's company first reduces the flakes of this mica-bearing mineral to nanoscale proportions—about $2 \times 10 \times 50$ nm. Then it embeds these nanoflakes evenly throughout a very thin butyl-rubber matrix. The result is a material that places trillions of tiny barriers in the path of any air molecule that wants to exfiltrate. As it is forced into an escape path with countless twists and turns, the air molecule is slowed considerably. The net result is 30–300 times improvement in air retention for a given barrier thickness, even under pressure. Here's another way of stating that figure: One millimeter of butyl rubber can be replaced by as little as one-hundredth that thickness (10 microns) of Air-D-Fence.

Not sound like much? Imagine vehicle tires whose air pressure could safely be checked only once a year. Whose higher operating pressures reduced rolling friction, minimized heat and wasted energy, and lightened the tire by up to 5 percent. Goldberg estimates that reconfiguring all vehicle tires in the continental United States to replace butyl with Air-D-Fence would indefinitely free up to three million gallons of gasoline per year. That's a lot of freedom.

This type of incrementalist approach, Goldberg says, is likely to prove the most successful business strategy for nanotech in the short run.

"Our clients aren't interested in nanocomposite materials per se," he says. "Like their own customers, they're interested in features—lower weight, longer air retention, lower cost, and the like. As long as these features are provided, nobody cares if that's achieved with black magic, nanotechnology, or something else."

Furthermore, Goldberg points out, this application approach earns money as soon as possible after initial R&D. "Incrementalist firms such as InMat can keep going when firms that say 'let's hit a home run once a decade' go under."

Besides vehicle tires and sports balls, InMat is also looking at diversifying into anticorrosion coatings, abrasion protection, and other applications. One highly topical possibility is homeland defense. InMat coatings would protect laboratory workers handling toxic chemical and biological agents, in labs that safeguard citizens against terrorist attacks.

"It's true nanotechnology," Goldberg says. "It goes far beyond the empirical use of naturally occurring nanoparticles. It requires a deep understanding of properties and events in the nanometer range. We routinely intervene at the nanoscale to diagnose and engineer our air-barrier coatings." How, might one ask? "We've already modified Air-D-Fence to make it more elastic. It now takes strains of up to 20 percent without damage. This [property] and its low, long-term impermeability makes it ideal for chemical and biological applications such as airtight suits and gloves. We can do this cost-effectively because from initial idea to marketed product, our product-cycle times are very short."

CHAPTER 4

NANOFORNIA

NOWHERE TO RUN

SMALL TIMES IS a trade magazine that covers both nanotechnology and MEMS (microelectro-mechanical and electrical systems). In April 2002 it identified six U.S. hot spots in nanotech. In ascending order of importance these were: Chicago, Dallas/Houston, New York City, Boston, and Southern California around Los Angeles (SoCal). Right atop the list, *ichi-ban, numero uno,* described by *Small Times* as having "youth, money, brains, and glimmerings of that gold rush spirit," was SoCal's glittering younger sister from NoCal: Silicon Valley.

This is a young place, but a fabled one. It starts in southeastern San Francisco and stretches down, mostly inland from the Pacific Ocean, toward Monterey. As late as 1960, America used to think of this area (when she thought about it at all) as John Steinbeck country. Back then that author was its greatest claim to fame. But however much America loves its writers, it loves its money more. As a nation, the United States reserves its highest accolades for business people—measuring success less as lives illuminated than as dollars accumulated. Hence the towering profile and long shadow of the Silicon Valley myth.

It was here that the integrated circuit first took shape. It's where the personal computer was conceived, carried, delivered, and placed on the doorstep of an unsuspecting world. From Larry Ellison's Oracle to Scott McNealy's Sun Microsystems, from ACLARA and IBM's Almaden Research Center to Xerox's Palo Alto Research Center (PARC) and the great concentrations of academic know-how at Leyland Stanford University, Silicon Valley was the flying saucer's landing pad: the place where the future came to earth.

Judging by the fervor with which NoCal is adopting nanoscience, it still is. It's no coincidence that Jeff Jacobsen, CEO of one the area's leading nano-tech firms, calls his company Alien Technology. For a recent magazine photo he posed with Gort, the foam-rubber robot from Michael Rennie's 1951 sci-fi film, *The Day the Earth Stood Still.* Alien (the company) applies a proprietary technique called fluidic self-assembly to make integrated circuits with features as small as 50 nanometers, which approaches the true nanoscale.

Silicon Valley still has "it"—though defining *it* evades most analysts. As well as world-leading research nodes, *it* includes enormous pools of venture capital and savvy people who direct it; experienced facilitators, analysts, and consultants; and thousands of firms with expertise applicable to nanotech.

Even more important are Silicon Valley's battalions of The Young And The Stupid. TYATS are energetic technical whiz kids who constantly perform miracles because they're too callow to know what can't be done, or else too enthusiastic to give up even when they do know. Interestingly enough, the dot-bomb of the late 1990s has only slowed, and not stopped, Silicon Valley's push into nanotech and nanoscience. The main effect of the recent meltdown in e-business has been to dump more of The Young And The Stupid onto the job market. These TYATS are hungrier than ever and willing to take a flyer on an emerging sector that's full of promise but, to date, still largely untried.

Most important of all to *it* is something that defies description. *It* is a charge, a feel, a mix. *It* shares something with great cooking, great poetry, and great marriages. "It's not the ingredients, but the recipe that counts," says local analyst Douglas Henton, referring to the rise of nanoscience in NoCal. "It's the culture."

Silicon Valley's culture is as unique as yogurt. It may be the only place on earth where failure is seen as a precious asset rather than a cause for shame. To the Valley, crashing a company is like completing a practicum in grad school. When you've experienced what doesn't work, you'll know what will. There's no better way to learn. "If they fumble," *Small Times* writer Candace Stuart observes, "they regroup and try again."

In May 2002 I flew to San José for Nanotech Planet, a floating conference on nanobusiness that later took off around the globe by way of London, Berlin, and Singapore. I went to log data and gather business cards; I ended by filling my lungs with *it*.

This caught me flat-footed, because I'd thought that technology had pretty much freed itself from place. That phone you're using may have parts from Mexico and Indonesia; the cop car that just flagged you down was put together in Ontario, though designed in Detroit; your database software might have been written in Germany, Pakistan, or Eire. But the process of disconnecting land and ideas isn't completed yet, and may never be. Certain regions still punch above their weight. When it came to NoCal, I found, a book that I'd initially planned around technical subsectors had to recognize that old devil geography.

Anthropologists call a locale like Silicon Valley a "culture area." The concept acknowledges that some things can't be packed up and exported along with software, hardware, personnel, and other bits and pieces of technology. *It* stays in Silicon Valley as the canals stay in Venice—and I don't mean Venice, California.

AFTER MY FLIGHT leaves Vancouver in driving rain and climbs into a mat of clouds, I don't see the sun for ninety minutes. But at the Oregon-California border, the low-pressure system stops as if it's run into a tough state law. The clouds vanish and the orchards, vineyards, and oak forests of NoCal stand shining in a clear cool sun. No sign of an economic slowdown here.

The fine weather continues into San José International. My hotel is just around the corner from the arrivals lounge. I check in at noon, change into running gear, go back downstairs, and ask a desk clerk where's the best long-distance route. He stares at me blankly.

"We have a fully equipped exercise room, sir."

"I know that. It's a gorgeous day and I want to run outside. Where do I go?"

"Well, sir..." He clears his throat. "I'm afraid there isn't any place."

Now it's my turn to stare. "What do you mean? What's *that?*" I gesture out beyond the windows where a sun-drenched, tree-lined avenue stretches north and south.

"The street ends two hundred yards in either direction," the clerk says. "Freeway south, no sidewalks, pedestrians not allowed. Construction site north. Sorry, sir. You wouldn't even have time to work up a sweat."

In the exercise room I find a huge, rawboned, redheaded man holding a gym bag. "You asked about routes?" he says glumly, and I nod. We run together in the workout room, logging miles together on treadmills, hiding from a perfect California day.

The car rules in Silicon Valley. For work, play, social life, and exercise, you drive from indoors to indoors. Cars in California have done what eluded the Nazis: They dominate their ecosystem by ruthlessly squeezing out every other form that could possibly compete. The auto is steed, enabler, companion, and status symbol. It's even a jokebook. A BMW plate in the hotel parking lot reads:

RCH4A★

Happily for me, my exile from the sunshine pays unexpected dividends. My fellow treadmiller turns out to be Dr. William L. Warren, a speaker at the conference. He's a Midwesterner, based in Stillwater, Oklahoma, and he has the virtues of the breed: big heart, big frame, big brain, and a skeptic's nose for B.S. Of everyone I meet here, he has the best sense of humor; in fact, he's one of the few who demonstrates a sense of humor at all. Like Rafael Sabatini's fictional hero Scaramouche, he's "born with the gift of laughter, and a sense that the world was mad."

There's a lot of madness in the world Bill Warren now proceeds to describe. As we exercise, Bill gives me an expert's take on the origins of U.S. nanotechnology. I know value when I see it, and Bill delivers pure gold: the smart, observant insider's point of view. As he talks, I take notes madly in my head.

First off, Bill's a contrarian. The previous administration, Bill tells me, pushed nanotech because Vice President Al Gore wanted to throw a bone to the physical sciences. "For the last two decades the biosciences have gotten disproportionately high funding from the [U.S.] government," he tells me. (I'm not deleting any gasps here; Bill's in better shape than I am.) "Gore sold nanotech to [U.S. President] Clinton, who made it a full federal program. When Gore didn't win [the 2000 U.S. presidential election], anyone who had been behind the nanotech program cut and ran from it. What they didn't figure on was that by then, too much time had gone by for anyone to shut the blasted thing down. It had too much momentum. People had already started doing research and spending money. All these congressmen and senators got screamed at from their constituents all over the country, demanding they continue the NNI [National Nanotechnology Initiative]. [President George W.] Bush was in a corner. He ended up not only keeping it but expanding it."

It's an old story. Whatever changes science and technology create together, one thing is constant: Hell hath no fury like one scientific discipline convinced that government funders are favoring another. Notwithstanding the hundreds of millions of new federal dollars pouring into the NNI, let alone the hundreds of billions given to physics, chemistry, engineering, and electronics from 1940 onwards by nuclear and conventional weapons systems plus the space program, the physical sciences in the United States continue to smart from the goodies given to bioscience over the last twenty years. It's true that an official War on Cancer, together with biotechnology's suddenly revealed abilities to deliver amazing new drugs and therapeutic protocols, gave bioscience a relative edge after 1980. But averaged over the last sixty years, biosci funding is not excessive. Back when physicists were the spoiled darlings of the feds, when even far-out theoretical thinkers like Dick Feynman landed college posts and federal grants as a matter of routine, biology was treated like the wheezy old guy who dusts butterfly cases at the museum.

Not that you'd realize any of this from the engineering rhetoric making the rounds today. "Only about 15 percent of the proposals competing for NNI funds last year were successful," said Hewlett-Packard's Dr. Stanley Williams in a speech in Palo Alto in May 2002. The NNI review committees, he complained, "turned down as many high-quality proposals as they funded....U.S. investment in fundamental research is out of balance, with the growth in biosciences larger than the community can sustain and stagnation in physical sciences and engineering choking out economically promising and important areas." *Daaaaaaad!! You gots Jimmy ice-cream and you didn't gots me nuffin!!!*

All this is a vast source of amusement for Scaramouche de Oklahoma, who is as clearheaded an individual as I've met.

The next area across which Bill Warren sweeps his mental spotlight is venture capital. "Don't want . . . to knock VCs," he tells me as we work the weights. (Okay, I confess: He works, I watch. This is one strong guy.) "VC myself. But when most ... of these people ... say 'due diligence' ...

they just mean ... *business* due diligence. ROI and debt load. Management capa ... "—he puffs—" ... bility, and so forth. That's important. But it's not ... the whole story. The minute ... most financial guys ... run up against ... science, they start ... behaving like ... some dip-brained deb. At a prom. 'You're from UCal? At Berkeley? Work in nanotech? Here! Take ten million!' You'd think the ... dot-com bubble ... had never happened. Ungh!" *Clang* go the weights as he sets them down. They're heavier than I am.

Scaramouche (sorry, Warren) speaks from experience. He got his engineering doctorate from Pennsylvania State in 1990, worked as a program manager for DARPA—the U.S. Defense Advanced Research Projects Agency—from 1998–2001, and is now president of Sciperio Inc., a VC firm with interests in ceramic and molecular electronics, water purification, and advanced manufacturing. While doing all this, he somehow found time to author 200 published papers, file five patents, organize five international conferences for major professional associations, and win more national awards than most of us have even heard about. (R&D 100 Award, 1997; Industry Week Innovation and Technology Award, 1997; *Discover* Award, 1998; Sandia National Laboratories Award for Excellence 1993, 1995, 1996, 1997, 2000...)

Bill Warren has configured his VC firm Sciperio in an unorthodox way. The firm acts as an administrative core, vetting new technologies and locating funding for them. But Sciperio does not itself undertake any commercial developments: It spins off new daughter companies for that.

I find a clue to Scaramouche's character in a key detail. While Bill's spin-offs seek and accept VC and angel funding and also go public, Sciperio won't take a dime from anyone, anytime, ever. I ask Bill about this.

"If you want a shot at megabucks, go public," he tells me. "If you want to stay true to your own vision, don't. You'll stay smaller, but you'll run your own show." He grins, nods, towels his neck, and lopes off down the hallway, "wild Hamlet with the features of Horatio."

SLOUCHING TOWARD BETHLEHEM

"There's a lot of promise in nanotechnology," Bill Warren told me in the workout room. "But I agree [with Harris Goldberg of InMat] that it will have to be incremental and market-driven at first. It will have to progress, and finally take over, by slow steps." Like solar power, maybe? Years and billions were spent on fancy ways to convert sunlight to electricity. Then a passive-solar innovation of idiot simplicity came along and saved more BTUs than the rest of its ivy-tower cousins put together. It's the solar blanket, which lets the sun replace oil, gas, or electric heaters to warm the water in a swimming pool.

"Even less apparent than your example," Bill says. "At least some consumers understood at some level that a *solar* blanket harnesses *solar* power. Nanotech's first steps will be all but invisible to product users."

If Bill's views on venture capitalists are accurate, those steps may even be invisible to some of their funders. I'm convinced of this by a long afternoon session on the conference's second day.

At the afternoon workshop session, VC after VC steps to the microphone and pleads for help from someone, anyone, to make sense of nanotechnology's technical terms and concepts. Even the VCs with solid backgrounds in established sciences seem at sea. The field is so new, its discoveries so groundbreaking, that it seems beyond any one individual, no matter how well educated, to keep up with it. Even VCs with doctorates in biotech have to pick up the phone and deal with a team in Iowa that produces nanoparticles biologically, but then wants to apply them to inorganic substances. On the margins of my notes I scrawl both Coleridge—*We were the first that ever burst / Into that silent sea*—and Yeats: *Who knows what rude beast, its hour come at last, / Slouches toward Bethlehem to be born?*

Nobody doubts the VCs will ultimately orient themselves. They did it for other scientific and technical revolutions—biotech in Boston and Montreal, and before that microelectronics right here in Silicon Valley. It will all come to a focus in three years at most; but for the moment, there's confusion. Certainly in funding, and probably even in R&D, this is nanotech's Wild West.

At the moment, it occurs to me that the mood in this room reflects both ideograms in the Chinese phrase for *crisis*: Danger + Opportunity. You can cut the excitement with a knife—and the uncertainty and greed with a chainsaw.

"Do you have a super project?" asks George Lee, a VC from Glimmerglass Ltd. in Menlo Park, California. "Do you have great management? Then go out and get a good VC. You're going to burn through capital at an amazing rate. People will watch how you do this, and watch what you accomplish as you go along. But it's the VC who'll come up with your money. Banks won't give you a dime unless you don't need it. They won't give you an unsecured loan, or rather they won't give you any loan that's secured only by IP [intellectual property]. Out of the goodness of their hearts the banks may offer you credit-card financing, with the low, low cost of 27 percent per annum. Stay away from that, if you can. Also avoid cute little accounting tricks like receivables factoring."

Voices from the floor: *What? What's that?* Ah, America. One would never encounter such bold, ungentlemanly interruptions in Lausanne. But this rudeness instantly clarifies the issue.

"Receivables factoring works like a ghetto cheque discounter," Lee answers the unidentified questioners who spoke for all. "You assign as-yet unpaid invoices to your money sources. And this, you think, will take care of capital repayment. But what happens if there's a technical glitch and your product shipments are delayed? You don't ship, you don't invoice, you miss your loan repayments. And then you're dead."

One after the other, the VCs paint a generally gloomy picture of capital flow for nanotech. Three or four years ago, all sorts of money was being thrown at the dot-coms. Now that sea of funds is drier than Death Valley.

"You need a thick skin to be in a start-up," Lee says. "You may break out the champagne at 5 P.M. Friday when you hear you have funding. At 8 A.M. Monday the phone rings: Your money source has rethought things and is pulling out. Tough on you." Voices in the audience groan word-less agreement: Lee has obviously touched a nerve. I shake my head at this. The VC community seems to have gone from thoughtless spending to an equally thoughtless penury. If five years ago they'd taken Bill Warren's approach and made their technical due diligence as rigorous as their nontechnical enquiries, they wouldn't have been so badly burned. But too often they confused a business plan with an IPO, and shares with products. Now they're shying away from good firms that have every-thing—ideas, people, technical head starts, proven technology, good markets, patented IP—and leaving excellent alliances at the altar.

Happily, that's not the end of nanotech. For into the financial breach left by business has come government. "Federal and state programs, even municipal ones, are currently the best place to go for seed money and early-stage funds for nanotech ventures," says Michael Fancher, a VC from New York. "They'll give you thin tranches"—that is, only a few bucks at a time—"but they are a [nanotech] start-up company's most likely source of initial funds. Remember that every state and municipality in existence is constitutionally committed to fostering its population's health and security. Tap into that, and you'll get your funding. Not megabucks, but tens of kilobucks. Anyway, more than zero. Which is more than what you'd have otherwise, right?"

Despite having to scrape by on nickels and dimes, nanotech ventures are generally expected by their funders to think big. "Don't be content with projecting mere survival," Fancher advises. "Aim to rule the world."

Rather like *Eurokrats* who aren't afraid to rejig a proposal before granting financial help, VCs—at least in the USA—may dictate details to firms that come seeking funds. A good VC can even organize a consortium, bringing together various companies and unifying multiple business plans into a single, more powerful one. In several cases, Fancher announces, aggregate funding has been quadrupled in this way. Dig it: the VC as CEO.

Despite these success stories, nanotechnology has a long, winding, uphill road to walk to reach the average VC's heart and pocketbook.

"Nanotech and MEMS both suffer from the same flaw," mutters a jaded East Coast VC who declines to be named. "They have no clear aim. Jesus Christ, they don't even have clear definitions of themselves. I mean, latex paint is full of nanoparticles. But is it nanotech?" He shrugs. "Who knows?"

Some VCs have different notions. Ajay Ramachandran, a general partner at the VC firm of Ark Venture Partners, has university degrees in both molecular cellular biology and biomathematics. He describes himself as "very bullish" on nanotech start-ups: "We like to get involved early, even at the back-of-the-napkin stage." Word of his receptiveness has got out, and now Ramachandran receives five to ten new business plans per week in nanotech.

"Unfortunately for us, most of these plans deal with new research," Ramachandran laments. "Few offer a workable technology that lets an investor such as Ark get involved." He cites a recent exception: a group of academics from UC Berkeley have developed a biomimetic (nature-mimicking) system based on how the human eye sees reality. The result: data compression protocols that are orders of magnitude more efficient than current alternatives. If these new algorithms continue to work out, says Ramachandran, "you may soon get full-motion video on a standard telephone handset."

A quick note on why data compression is important. As anyone with a computer is aware, visual information gobbles up a lot of memory. The continent-wide shift to high-speed Internet connection largely stems from images' insatiable appetite for bytes. A picture may be worth a thousand words, but it takes up a hundred thousand times the disk space of mere text.

Engineers have found ways to crunch visual information so that it places less load on IT hardware. If you're sending video, you can transmit each successive still picture by specifying only what's changed since the still that came just before it. This is called a compression algorithm.

But as science is starting to understand, the human body has the cutest IT tricks beat six ways from Sunday. For example, our eye-brain system has something called an orientation reflex. This briefly speeds perception when we shift our gaze, especially if we're alarmed or startled. If you've ever glanced at the office clock and thought its second hand had frozen, that's the orientation reflex. In emergencies, it appears to slow down time.

Duplicating such elegant efficiency in an artificial system, thereby letting it adapt effortlessly to changing circumstance, would mark a massive advance in visual IT. Once again, it's clear that the most important thing we have to do in exploring the nanocosm is to shut up and learn.

Like almost everyone who has anything to do with nanotechnology, Ramachandran has his own definition of the field: "It's technology that lets us manipulate atoms and molecules to create a salable product." In his view of things, nature rules: "If we duplicate only ten percent of nature's skills and powers, we will revolutionize every industry in the world. We're starting to find this now in the plans and proposals we see, especially in food, shelter, and textiles."

Despite his optimism, Ramachandran fears a nano-bubble is forming. "Hype is building without sufficient understanding of technology from the bottom up. Yes, we may see nanoscale supercomputers, medical nanobots, amazing inventions that usher in a whole new age. That's all possible, sooner or later. But let's temper our enthusiasm for this vision with some pragmatism and skepticism. The fact is, nanotechnology is still nascent."

Ramachandran predicts as follows: Through 2004, expect an enablement stage of new discoveries in basic nanoscience. From 2004–2008, expect a development stage of emerging technologies. And from 2008 onward, look for a surge of mass production in nanomaterials, biomimetic software modeling, and nanosurgery.

"We can," he says, "already model new substances at the nanoscale using CAD [computer-aided design]. I foresee a trillion-dollar market for nanotechnology by 2012." He pauses, frowns, shuffles notes, and sighs into the podium microphone. "Of course, all this assumes that VCs can be got up to speed on the science and technology behind nanotech. There's a real need to educate investors. At the moment, they find the whole area far too complex for comfort."

At lunch the next day I find myself in the midst of a representative population for this conference. To my left sit, in order of distance, a middle-aged academic doing research in basic nanoscience at Hunter College of New York City University; an engaging late-twenties woman with an M.Eng. degree, looking for an entrepreneurial opportunity in nanotech; and a beefy older man with a master's degree in aeronautics who is CEO of a midsize firm producing nanoparticles. To my right is a young bearded man who proves to be the U.S. correspondent for a Parisian business paper, and an Asian-American gentleman, impeccably dressed, who listens much and says little.

The conversation turns to VC-bashing. Capital is squeezing new ideas to death, the correspondent observes. Five years ago it was so lax that it helped create the dot-bomb. Now it's as bad as the banks: It won't make a loan until it's convinced you don't need it. I'm sorry to hear that, says the woman engineer—I'll need start-up capital as soon as I find the right technology, but I hope I don't run into a VC like the ones you're talking about. Better hope you don't, says the correspondent: He'll eat you alive. "He'll take your intellectual property and spit you out like a grapeskin." I contribute a phrase from *Rats, Lice and History* by Hans Zinsser, a microbiologist writing in 1934: "The cow eats the plant. Man eats both of them, and bacteria (or investment bankers) eat the man."

Into the laughter comes a soft comment from the Asian-American gentleman:

"I'm a venture capitalist."

Dead silence. Oops.

"Your comments refer to a certain type of VC," the man goes on. "The investor at the B or C level. So your remarks, up to a point, are valid. But the A-level venture capitalist does not want to trash a company for a quick return. He, or she, will commit for at least the intermediate term and most likely the long term. Perhaps even the very long term; that is, indefinitely. That type of VC is your best friend. Any firm would benefit from such an investor."

Right. Chastened like a bunch of schoolchildren, we rise and go off to attend the next meeting. Guy had a point.

GLORY, JEST, AND RIDDLE

At times this is a strange, uneven conference. Two phrases keep popping into my head. One is from fellow Vancouverite and sci-fi author William Gibson: *consensual hallucination.* Gibson was talking about cyberspace, but boy, does the term apply here. If nanotech ever becomes the juggernaut its advocates predict, it will be not just through intrinsic excellence, but also through self-fulfilling prophecy. All these scientists, engineers, analysts, bankers, venture capitalists, and other highly trained professionals will simply have agreed (in Jean-Luc Picard's famous phrase) to "make it so"—and it will happen. Since this is how we see things, thus it will be. Physics, if necessary, be damned: We can always invent new physics. Where there's a will, there's a way.

The second phrase that keeps intruding into my thoughts is Alexander Pope's sad, incompressible summa of humanity: *The glory, jest, and riddle of the world.* This week I'm rubbing shoulders with as wild a variety of sentient creatures as exists beyond a diplomatic reception in *Star Trek.* There are geniuses, charlatans, and people who seem to be some of each. There are superb science, solid business, great ideas; and then there's the other stuff. I meet a Caucasian analyst with a long gray ponytail who presses palms and bows instead of shaking hands. He talks about "shaping future history," and his handouts contain gems like this one:

> Synthetic brain engines, which themselves are nano-manufactured synthetic organism components, become interconnected into the ubiquitous process brokeraging operational ecology, which in turn develops into synthetic sentience process organelles evolving into global scale macro entities.

He calls this "thinking out of the box"; I call it "thinking out of the brain."

I also meet a wild-eyed Ph.D. from Columbia who treats his gentle wife with savage contempt and snarls at her whenever she tries to interject a comment into his endless harangues. In his lab, he cages atoms and examines them in isolation; evidently he does the same thing with his wife.

Yet another champion has sectioned a spider's eye "down to a resolution of Planck 7" (uh?) and "proven" that a completed electrical circuit exists, with energy leaping out of his micrographs like volcanic plumes from what are obviously cellular parabolic transceivers. What it proves is that the brain, of course, is a mere bit-shuffler, devoid of creative thought . . . .

So tell me, sir: If brains cannot think, how did your brain come up with its profound insights? Don't tell me, let me guess. In this color photomicrotomograph, which I just happen to have stored in my computer, we see Yahweh (fig. A) delivering the performance specs of the angelic realm (fig. B) to the archangel Gabriel (fig. C), who turns them into hard spec (fig. D, lower) and reads them aloud to yours truly . . . .

Did the silicon revolution saddle itself with such goofballs in its messianic stage, thirty years ago? There are times when nanotechnology seems to draw its personnel from St. Louis, Missouri Territory, circa 1849. Here are drunkards, preachers, hopeful innocents on their way to the Sutter's Fort goldfields, U.S. Cavalry troopers, and a virtual cavalcade of snake-oil salesmen. Up in my room that night I type this note: *If this motley brigade manages to discredit their whole newborn discipline and bring it crashing down, nanotechnology may yet prove to be the best idea that never made it out of Palo Alto.*

As I am typing, an earthquake strikes, as if on cue. I diagnose it instantly as the P-waves pass: midrange, shallow focus, epicenter fifty miles away. Jules Verne said of his hero Hector Servadac that if he were shot from a cannon, he would spend his last seconds mentally computing his own trajectory. I read that at age 11 and decided it was cool to think like that; I still do.

The glasses in my bathroom are clattering merrily. The room doesn't seem to be moving, but that's because I'm moving with it. My inner ear tells a truer tale of roller-coaster movement: up, down, back, front, sideways, and repeat. I glance out the window; the water in the swimming pool is sloshing around. But the electricity doesn't flicker, and trucks and cars tear along the freeway without slowing. The life form that conquered California isn't going to stop for a mere earth tremor.

BACK TO ANALOG

Tom Theis believes in analog data processing. He thinks it's the future of computation.

At the moment, nobody else in the room appears to notice the revolutionary nature of this statement. Although they're on the edge of their seats for this keynote address, they're too involved with other agendas, which they flatter themselves are hidden, to actually audit the content. They're here to adore or condemn: to kiss ass or kick it. This is because Dr. Thomas N. Theis isn't just another speaker at this conference. He's director of physical sciences at the Thomas J. Watson Research Center in Yorktown Heights, New York. As such, he's a big gun at the biggest gun of all, the Big Bertha of knowledge firms: International Business Machines Corporation of Armonk, New York.

Some statistics: IBM's gross revenue in fiscal 2001–2002 was almost U.S. $90 billion. That's greater than the yearly federal budgets of half the world's nations. Among multinational corporations, IBM takes a back seat only to a handful of monsters like Shell Oil, ExxonMobil, GlaxoSmithKline, and General Motors. IBM's global workforce has declined from its 1986 high of 407,000—a figure approaching the population of Greater Denver—and now stands at a mere 300,000. Still, that's as if every man, woman, child, tourist, and visiting stockbroker in Staten Island, New York, and three-fifths of its wharf rats, worked for IBM worldwide.

Within this corporate juggernaut, a singular culture has arisen and continues to perpetuate itself. In the 1950s, IBM employees called themselves "EyeBeeYemmers." Back then they ate at IBM cafeterias, played golf on IBM courses, vacationed at IBM resorts, and gathered at morning meetings to sing official IBM songs. All together now, *alla marcia,* to the tune of "Jingle Bells":

> IBM, happy men,
> Smiling all the way,
> Oh what fun it is to sell
> Our products night and day!

IBM, Watson men,
Partners of T. J.,
In his service to mankind—
That's why we are so gay! (*sic*)

The 1954-model EyeBeeYemmer was the original Man (men only) in the Gray Flannel Suit. A lot of this has eased and softened; but there remain certain . . . rules. Nylons for women, seamless. Sandalfoot is possible, but no sandals, even in summer. Power two-piece suits, or sensible dresses with high necklines. Bare shoulders? Forget it: Full sleeves are required. For men, white shirts whose ties are kept as they should be (well knotted) and where they should be (snug to the collar button).

To compensate for these strict standards, or perhaps to enforce them, temperatures within IBM buildings are kept, well, *brisk*. Whether it's 90° F or 70° F outside, offices will be maintained at a thought-provoking 68 degrees. Don't expect what engineers call a "floating HVAC delta-T"— that is, an internal temperature that rises or falls with the outside temperature, keeping a fixed distance between the two. When it's 68 degrees outside, your physical transition to the IBM corporate culture will be so smooth it's unnoticeable. When it's 98 degrees, your system will take the Polar Bear Plunge. (This is something done in certain Canadian cities on New Year's Day, when manly men and other morons wade into seawater the temperature of melting ice. Inland, where the water's fresh, they saw holes *through* the ice and take a bracing dip in salt-free liquid of identical temperature. That is the general effect of encountering IBM air conditioning on a hot day.)

EyeBeeYemmers are expected not merely to adapt to this air-conditioned assault; they are expected not to notice it. To wish away blue noses and shaking frames in mid-August; to be in denial about their firm's technological Ragnarok, the dreaded winter-in-summer that in Norse myth signals the end of the world. Never mind the wolves come a-ravening or the snowdrifts deep in the corridors, or the suggestion that Big Blue derives its name from the color of its workers' skins. (Although that's proved to be an excellent business strategy. With everyone a nice deep hypothermic indigo, there are no visible minorities and hence no need for affirmative action.)

No, it's none of that. Sixty-eight degrees Fahrenheit is good for the machines, that's all. This Arctic temperature is also good for people who want to *work* like machines—on call whatever the time or the day, constant in their productivity, and taking downtime only when it's minimal and prescheduled. Sorry, Jones, you can't vomit today: We haven't planned for it.

The oddest thing about this odd culture—mechanized, reified, regimented, whose motto is *Think*, yet delivers so much that is totally prethought—is that it works. Somehow, God knoweth how, team spirit ignites and develops at IBM; new ideas germinate and prosper. There are those who would deny this, just as there are EyeBeeYemmers who would die under torture rather than admit their basic freedoms are curtailed. But the facts speak for themselves. IBM has made a difference to the world economy, not only through its size but with its brains.

The story behind IBM culture makes for vivid reading. IBM is a vertically integrated company: that is, its activity goes from theory to manufacturing and covers every product and process in between. At the top end, IBM undertakes basic, curiosity-oriented research into the behavior of matter and systems. The company identifies early-stage ideas that emerge from this research and nurtures them into prototypes. If the ideas continue to work out, they are ultimately embodied in salable products.

This *modus operandi* makes IBM atypical, at least compared to firms that are currently its equal in size and corporate revenue. The IBM approach is historically mainstream; it is typical for a big company two generations ago. Others may change with changing circumstances, but not IBM.

This continuity of culture occasionally makes IBM—known everywhere as Big Blue for its corporate color (i.e., the hue of its frozen employees)—an object of ridicule. Thousands of other U.S. corporations, based on the cold analytical logic of profit and loss, have reduced, outsourced, or flat shut down their in-house R&D operations. But to its credit, IBM has successfully resisted this holy grail of the quarterly P&L accountants. In so doing, the company has demonstrated the reasons for its past market dominance and laid the foundation for its continued supremacy in the future.

It hasn't been easy for IBM to buck American industry's stampede away from long-term R&D. Blame Big Blue's hidebound intransigence or laud it for sticking to a noble vision; but at times the company has seemed almost alone in retaining well-funded laboratories that do curiosity-based research. In my estimation, outside of the multinational pharmaceutical giants, only Hewlett-Packard and Microsoft Corp. rival IBM in this.

IBM's achievement in continuing to support basic R&D reveals its scope, its Medici-like grandeur, only when you put it in the broader context of technology transfer. That's a study in itself, and deserves a sidebar.

THE INNOVATION CURVE

If you graph things on a logarithmic scale, with time increasing as you moved rightward along the X-axis, you end up with an S-shaped curve that looks like a ski hill, with the mountain on the right (Figure 4-1). This curve summarizes the expense of inventing, developing, and establishing a new technology. The ground starts off level, but as you move to the right, the slope steepens until it would daunt an Olympic champion.

Here's the interpretation. At first there's just a brainstorm, an idea. Ideas are not totally random. You need some background in whatever field your concept applies to. Nonetheless a technical idea is like lightning: It can hit anyone, anywhere, anytime. In business terms, many a new idea is effectively free. This fact maddens firms that invest millions of dollars in formal research teams and are then blown out of the water by some TYATS upstart operating from his mother's front room.

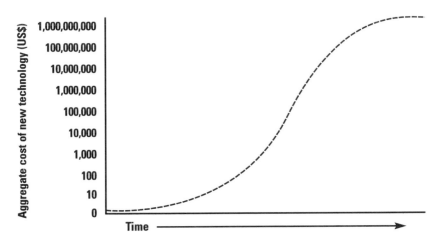

Figure 4-1. Innovation Curve

As time goes on, the innovation curve steepens. The first step involves widening your scope away from a specific problem and scrutinizing the wider business context. So you've thought of a better way to connect aluminum car parts on high-speed robotic assembly lines. Are there other ways of doing this? What are today's dominant technologies? Who owns the IP? Answering such questions requires smart people with good qualifications, and that costs money. From this point on your idea ceases to be free; henceforth every additional inch up Mount Innovation is more of a financial slog.

Assume the concept works out—that is, the world has a crying need for your new technique. You still have to demonstrate that your idea works consistently, first on the benchtop and then on a progressively larger scale. Count on a cost increase of one order of magnitude ($10\times$) per step. Over three steps, expenses increase a thousandfold. By this time you the inventor have likely exhausted your private capital and must find external sources.

This ushers in a whole new level of pain. Now you must satisfy not just yourself, your technical colleagues, and your mother that your precious concept is workable; you must also placate a parade of bankers, venture capitalists, and other sources of cash. You must convince them that your idea is likely to provide a return on their investment high enough and sustained enough to justify their capital outlay.

As an alternative, a small firm created to develop a new idea may form an alliance with a larger company. Here the mouse trades her technological excellence for the elephant's production capacity, marketing profile, deep pockets, and sales connections. It's a tough balancing act for a mouse and elephant to manage a joint scale-up, but it can be done to the benefit of both.

If this project approach fails but the technology still seems appealing, the larger partner often buys out the smaller one. A company acquiring technology in this way is generally big, with significant cash reserves and access to credit. This is necessary because the final stage of innovation, the movement of new technology to the marketplace, is the costliest part of the process—the hardest part of the hill. If that idea for aluminum car parts (the *concept*) can be made to work consistently in progressively larger scale-ups (*proofs of concept*), then it is a valid technical advance (*innovation*). But for this innovation to be brought to market, with all the desirable positive effects on company earnings, market share, and stock price, one last stage will have to take place: *technology transfer.*

To be successful, technology transfer must address problems in mass production, customer service, positioning, and distribution. Until all such problems are solved, no technical advance—no matter how good it looks on paper, or how well it has worked in trials—can be considered complete.

In my book *Prototype* (Toronto: Thomas Allen, 2001), I examined forty operations that had successfully scaled the innovation curve from plain to summit. Except for a few cautionary examples—okay, horror stories—I concentrated on successes. As I had expected, given my experience at the National Research Council, almost all the firms I profiled got through their progressive business stages as I've just outlined. It didn't matter whether they were in manufacturing, like Martin Yachts; or manufacturing support, like Virtek Lasers; or agribusiness/ biotech, like AgrEvo. In every case, bright ideas began free but quickly turned expensive.

Now here's the kicker. While subject area didn't bear on innovation, company size did. This happened in two ways. One option led to havoc, the other to success. The second-biggest source of trouble in the innovation process came when a small firm thought it could do everything a big firm could. That's understandable. What was surprising to me was that the biggest glitches came when a big firm tried to act small. There are separate niches for elephants and mice, a fact that either one forgets at his peril.

Here's the key, I found: Big firms, at least the best of them, excel at technology transfer. But the small companies and individuals—the little guys—are the ones who *find* the new ideas in the first place. Small firms understand this, because years of success and billions of dollars have not made them (as a professor at Harvard Business School described her university to me) humility-challenged. Seldom will you find a little firm trying to throw its weight around; it's the large firms that get confused. For every mouse that tries to roar, twenty elephants attempt to squeak.

But, says the big company, surely we can do everything some squirt of a firm can do, plus a lot more? Theoretically yes; practically, no—or at least not often. As readers of *Cosmopolitan* magazine will attest, size does matter. There are things that a big firm—at least the majority of big firms—may do well in theory that seldom work in practice.

Admittedly, large firms have the HR depth and access to funding that lets them scale the curve's steeper slopes. This is what allows a big telco such as AT&T to connect entire urban areas to fiber-optic broadband and will someday let big automakers like Toyota and GM change over from Otto-cycle engines to cleaner-burning fuel cells. Such activity is immensely complex and a huge drain on corporate resources. Surely, then, a mountaineer with the strength and skill to scale the upper reaches of the innovation curve will have no trouble with the same curve's flatter, less demanding parts? A big firm that swiftly and effectively moves ideas to market should have little trouble coming up with those new ideas *de novo*, right?

Wrong. When I began my book research, my initial premise was that the bulk of new technology would originate in large companies, rather than in small and medium enterprises. To my surprise, I found I was in error. The technical ideas that most big firms were transferring originated outside those big companies. On the whole, the big players were technological trucking outfits, polishing other people's brainstorms and shuttling them from place to place. This got the ideas to customers with reasonable efficiency. But the ideas themselves originated less from the big firms' R&D departments than from small firms whose IP had been leased. Sometimes the whole firm was bought outright: lock, stock, barrel, staff, and garage-door opener. The brainstorm per se, as distinct from the process of moving it to market, first came in a lone engineer doodling, or two guys and their wives working together around the kitchen table.

I believe this is an almost inevitable consequence of corporate structure. As firms grow, and especially if they grow quickly, they are forced along a strict progression. Informality becomes formality; handshake gives way to written contract; policies and procedures dominate everything, including R&D. This tends to stifle innovation within the monster firm—but not elsewhere. Individuals prevented from thinking, dreaming, and experimenting at a big company simply strike out on their own, either by choice or by layoff. Once back in the figurative garage, they find their creativity flowing faster than ever.

To recap: Ideas are, like a theologian's concept of an angel, dimensionless and immaterial. Anybody can get The Big Thought—even a twentysomething patent clerk in Switzerland whose name was Albert Einstein. Great ideas are as likely to descend to earth in Nantucket as in New York.

Paulin Laberge, president of Altus Solutions of Burnaby, British Columbia, neatly summed up for me why no one has to be impressed by size alone.

"Everyone in this region's high-tech sector," he told me, "owes BCTel a vote of thanks. If they hadn't been shortsighted enough to shut down MPR Teltech [the BCTel R&D operation] and put us all on the street, we'd never have come up with so many profitable ideas." This was corroborated for me by a project manager laid off by Nortel and working for a ten-person company outside Detroit. "I don't have to spend all day making PowerPoint presentations," he said. "I can actually innovate, and I do." It's the flat, lower-cost part of the slope, granted; but that's where the journey begins—and that's where tomorrow's giant firms are born.

After months of interviews, I found that tiny details could accurately indicate where a firm lay on the innovation curve. On my cross-continental voyaging I'd hit a new urban area, confirm prearranged appointments, and then start cold-calling for local viewpoints. Inevitably, the bigger a firm was, the more hostile my reception. I came to realize this was because the big guys guarded information like heart's blood—not just in-house IP, but *everything*.

In seeking information, I was about as welcome as a borer to a corn farmer. Senior management was difficult to reach, middle managers yet more difficult, line engineers all but impossible. PR officials stonewalled contacts rather than facilitating. Employees were, as a rule, apprised that anyone talking to external media (me) could and would be summarily fired.

The typical large company, I concluded, sees the world as essentially hostile—holding more threat than opportunity. And since information is lifeblood, the only way to keep it from leaking away and bleeding to death is to wrap yourself in a data-proof barrier. Yet every firm that does this also prevents new information from getting *inside* to nourish, inform, and inspire it. The big firm wants a fortress but ends up building its own jail. It's not just cost that soars on innovation's S-curve. It's paranoia.

The contrast with the smaller firms I talked to was like night with day. In Ottawa, one of my cold calls was Tundra Semiconductor. Tundra is a fabless semiconductor firm that designs complex silicon chips with millions of integrated components for manufacture by offshore jobbers. Here's a summary of Tundra's instant response to my cold-call query: *Of course we've read your columns. We'll rearrange things to accommodate you. Would you like to interview our CEO, CFO, COO, or all three? No need to sign an NDA* [nondisclosure agreement].

Tundra Semiconductor proved to be one of the most innovative firms I profiled in my book. Not surprisingly, it was also one of the most open—and one of the smallest as well.

In growing, a firm consciously chooses to forgo chaos. But in embracing order, predictability, and settled form, it often kills the very conditions that once lent creative ferment. Formalization, however necessary to bigger firms, affects creativity like hardening of the arteries. This is why so many big firms lack intellectual adaptability, an openness to truly novel approaches, and the priceless talent of turning on a dime. Those attributes remain the core excellence of the smaller organization. There's an ancient example for all this. In an eye blink of geological time the big, slow dinosaurs gave way to small, fast, intelligent mammals. The dinosaurs seemed to have all the cards—but they couldn't adapt, and so they couldn't compete.

Now, these are only tendencies, not hard and fast laws. I'm the first to admit that there exist many examples of good new technology originating in large corporations. 3M invented Post-it Notes, a convenience for us and a moneymaker for them. Dow Corning produced Corelle, a microstructurally modified glass with excellent cooking characteristics. Yet even in large-firm innovation, the independent thinker, the intra-corporate maverick, is the typical source of ideas. Post-it wasn't a consumer need identified by market research: It was entirely "inventor push"—a bright idea from a 3M engineer who had to sneak around and conduct his initial research in near-secrecy. Corelle was produced by fluke when a curing oven malfunctioned, and a research scientist had the smarts to analyze the "ruined" product.

This being said, there are ways for a big firm to maximize its likelihood of finding new ideas. The trick is: *Act and think small,* whatever your corporate size. To foster innovation on the flat part of the curve, set up corporate microclimates that mimic conditions inside the smaller, funkier firms. Don't eliminate accountability; keep all your idea teams on the hook to produce. But clear away the small stuff—ritual meetings, dress codes, the punch clock that's actual or implied. And when your people start to produce, don't fire them for suggesting you stop producing buggy whips. Gold, including the allotrope called creativity, doesn't always appear in a convenient time, form, or place. It's where you find it.

- -

This lesson is, to its credit, one that IBM has realized. It might be more accurate to say the lesson was learned fifty years ago and never forgotten. In this way, IBM has beat the odds and confounded its detractors. It hasn't tried to be a mouse, nimble and alert; it's gone with its strengths. These include vast resources of money and people, (relatively) patient shareholders, and the ability to think—*Think!*—beyond the next quarter. But the last and greatest of its virtues has been, and continues to be, a functioning anachronism: a belief in fundamental research, even if what it discovers may not enter the market for twenty years or more. This great a depth of focus, this long a vision, still pales beside the 300-year business plans of big Japanese firms such as Hitachi or Mitsubishi. But for smash-and-build America, it's the equivalent of Plymouth Rock.

WE NOW TAKE YOU to Hall B of the DoubleTree Gateway in San José, California, where Dr. Thomas Theis is announcing IBM's official position on nanotechnology.

Big Blue is constantly sensitive to an emerging commercial consensus, whether among its customers or its competitors. Over the years it has displayed a genius for running around to the front of an existing parade and taking it over. IBM, Theis announces, has a goal: to remake its microtechnology division into a nanotechnology division. Theis accepts the emerging standard: For commercial as well as scientific purposes, the nanocosm ranges between one and a hundred nanometers. That, he says, is the length at which size really matters.

"Below 100 nanometers, the electron senses its quantum confinement and regular electronics ceases to work," Theis says. He now plays that remarkable Big Blue trump card: a depth of original research, applicable to nanotech, that stretches back even before Eric Drexler coined the word nanotechnology. The IBM milestones come rumbling out. The first scanning probe microscope, the scanning tunneling microscope, or STM, invented at an IBM Europe basic-research lab in 1982. The AFM, or atomic force microscope, ditto: 1986. Seizing, translating, and redepositing individual atoms: 1990. First intramolecular logic circuit: 2001.

"Note how old some of this stuff is," Theis says. (The room is still and silent; every eye in the audience is fixed on him.) "In fifteen or twenty years, the science we're doing today will be just as commercially important as those older discoveries are now." Nanotechnology, as Theis sees it, is "a long-term game . . . . It can take decades for concepts to move from lab science to products." But despite the time lag, it's an era that's fabulously exciting: "There never has been, or will be, a time like this in the whole history of science." Part of the reason for this is the ability of physical science to slow or reverse its recent relative decline in federal funding. (*Daaaaad!!!!!*)

"I'm enormously enthusiastic," Theis says. "But as people who are in some way involved in nanotechnology, everyone here today must take steps to manage the hype. I'm worried that perception and expectation are getting far ahead of reality. When bubbles burst, there are tears.

"Not," he adds, "that I'm against bubbles." The room, located at the exact financial, geographical, and spiritual center of Silicon Valley, erupts in laughter. "It says much about our society," explains Theis when things quiet down, "that it permits and encourages experimentation."

Theis proceeds to review IBM research. The silicon transistor, he says, is "already becoming a nanodevice." But silicon's unit hardware can't and won't get much smaller. "We're at the dimensions where the devices won't function if they shrink any more. Physics won't permit it."

As a possible alternative to silicon transistors at the nanoscale, Theis shows us a slide of the FinFET, a field effect transistor with a heat-radiating appendage sticking out of it. This fin is only 20 nm thick. This qualifies it as nanotech by Theis's accepted definition. In fact, nanotechnology has already begun to creep into computer hardware almost unnoticed. State-of-the-art heads for hard-disk drives have layers that are laid down on their surfaces with atomic-level control and tolerances of only ten nanometers. Or consider "pixie dust"—the engineers' nickname for a one-atom-thick rhenium layer that, applied to a drive head, leads to higher data retention at room temperature.

Theis's own lab has achieved data-storage densities of one terabit per square inch. I don't recall if the medieval scholasticists ever agreed how many angels could stand on the head of a pin, but a trillion zero-one distinctions can now demonstrably fit on a human adult's thumbnail. The device that makes this possible is the IBM Millipede, which uses over a thousand individual AFM tips to simultaneously inscribe ROM data on a hard substrate. Theis touts Millipede, as well as other technologies such as standing-wave electronics that have the same function as Millipede, as ushering in a new age.

"These things open up entirely new markets," he says. "They aren't really about data-storage densities. They're about incredible new things."

A skeptical thought crosses my mind. It may be true that Millipede technology could squash the Library of Congress into a wristwatch: We'll see. But at base, the incising technique itself is hardly new. In ancient Sumer and Akkad circa 3500 B.C., scribes used pointed styluses to cut symbols in wet clay. These symbols functioned as data outputs when irradiated by a noncoherent, nonpoint source of photonic wavefronts in the 1-micron range—namely, sunlight. Five thousand years later, the

Millipede reproduces this truly ground-breaking technique stroke for stroke. *Sumer is icumen in.*

Tom Theis thinks that 1 Tb/in^2 is about the limit for data density that existing technology can attain. Yet "existing technology" is changing as we watch. "Things are accelerating," he says. "The newest GameBoy"— a portable games console from the Japanese electronics giant Nintendo— "has a faster central processing unit than a personal computer with an Intel Pentium IV chip. We're approaching the end of Moore's Law."

That "law," first propounded by IT engineer Gordon Moore about forty years ago, states that computing power per unit area doubles every eighteen months. Another way of stating Moore's Law is that the cost of a given amount of computing power is sliced in half every 18 months.

Moore's Law, Theis thinks, may have another "ten, fifteen, even twenty years yet to go. But silicon-based technology can't go on forever." In other words, if Moore's law is to hold, then at some point in the next few years, something must replace silicon semiconductors. Theis thinks that something is carbon-nanotube technology. (See Chapter 8.)

"These things are amazing," Theis says, referring to the hollow cylinders of pure carbon, fifteen angstroms across, known as buckytubes. "Keep the covalent bonds straight, and they conduct electrons like metals. Twist the bonds, and they become semiconductors. But don't believe any claims you hear about buckytubes revolutionizing information technology in a few short years." For one thing, a buckytube transistor requires far more power to modulate than a silicon-chip microtransistor. "Besides, you'd need ten-to-the-twelfth [10^{12}] carbon-based transistors on a single chip. That's a trillion—one followed by twelve zeroes. To date, the record number for adjacent carbon-based nanotransistors is all of two."

Successful innovations, Theis says, tend to adhere to a pattern: They modify only one "business layer" up and down. Theis defines a business layer as the level or scope of business activity directly adjacent to, and immediately influenced by, an innovation. *Think small* is Theis's message. Modify the status quo, but don't try to explode it because you can't. You'll make money with an improved car tire, but you'll go broke if you try to replace all cars with hovercraft.

If miniaturization of components to the angstrom level is one goal of nanotechnology, Theis tells us, another central goal is self-assembly. It may be possible to mill, plane, and mold matter at the atomic level, but it would be much more elegant to persuade nanoscale objects to put themselves together.

Great idea.

Let's do it.

Right.

Ah . . . How?

"You need two kinds of information to build a snowflake," Theis suggests. "The first is in a tiny dust particle. This tells impinging water molecules how to minimize systemic energy. They use the dust particle as a matrix on which to self-assemble. Yet even given this *a priori* condition, you won't get any new self-assembled object unless ambient conditions are also right. In other words, you also need environmental information.

"Right now at IBM Labs we're providing both informational sets and getting honest-to-God self-assembly. We can create complex patterns and amazingly regular arrays."

At this point, as pants the hunted deer for cooling streams, the audience is hanging on Tom Theis's every word. And then, right in the middle of his latest revelation, roughly between *Thou shalt not commit adultery* and *Thou shalt do no murder,* a cell phone goes off. It's not a simple ring, either: no self-effacing beep-pause-beep. This thing has been programmed. It proceeds to play the first fifty bars of *Eine Kleine Nachtmusik,* followed by the French National Anthem, all at 120 dB—like an AirBus taking off a hundred feet away. Waves of hate spill out and break against the hapless bastard who sits jabbing buttons at his micro-anarchist, unable to shut it up. Finally he cuts and runs. Tom Theis resumes as the hall doors slam on the techno-fugitive.

"Self-assembly can occur with unusual metal alloys such as silicon-germanium or iron-plutonium," Theis says. "But it need not be limited to such exotic materials. Under the right circumstances, with necessary information input both *a priori* and from the environment, many physical systems will exhibit self-assembly. In fact, maybe most systems will."

Self-assembly is certainly not one of IBM's core businesses, Theis admits. "Nonetheless, this type of self-assembly process has been patented and looks very promising for future manufacture."

Theis pauses as if consulting his notes, except he doesn't have any notes. Everything's from memory, yet there isn't an *um* or an *ah* in his speech. This guy could steal a crowd from Alcibiades in the Athenian agora.

And his next remarks make my hands shake as I take notes. This is it: The Revolution. After sixty years of digital dominance, IT's central paradigm is on the brink of reverting to a primitive, long-abandoned, long-despised state. It is as if Big Blue's very hue is about to change to Big Red. The bombshell: Digital may soon be going analog.

Theis sets off his H-bomb with a dry theoretical query: How much digital information is necessary to specify any given structure? It seems a simple question, but under close analysis it gets insidiously complex. Things are simple as long as you restrict your discourse to microcomponents. Yet compare hardware to the living world, only for the briefest instant, and your whole logical structure breaks down.

To begin, consider the most complex synthetic artifact in IT—the modern microchip. The microcomponent exists on a plane. These planes may be stacked nine deep, yet they remain flat surfaces, not true 3-D. Nature, on the other hand, is totally tridimensional. (Theis's tone is dirge-like, almost lamentational, when he says this, as if he's reciting the sins of the world. The man is confessing to ignorance.)

And then there's the matter of data storage. "Our current technology needs tens of gigabytes of data to specify a video file," Theis intones. "Yet nature needs only three gigabytes to specify a human being." Here he's referring to the 3,000,000,000 nucleotides that encode the human genome. "*Something* is out of whack here. Obviously our set of IT algorithms, which is to say our whole conceptual understanding, is lacking. Under current modes of storing data, to write a file specifying even a simple living organism such as a paramecium would create a file that was unimaginably huge. Yet nature does it effortlessly, and in less space than a pinpoint."

Yes, we can store and manipulate data using digital electronics. But— and here Tom Theis, staid R&D director in a company that defines staid, holds clenched fists to the ceiling and shouts—"*We're just lousy at it!*"

When we look at life, Theis tells us—another cell phone goes off, *Das Ring der Californiungen,* and like its predecessor evokes breaking waves of loathing—we are compelled to be humble. (My God, it's about time somebody at this conference said that.) Living things store information in 3-D at the atomic scale, something of which humankind has only recently begun to dream. The simplest organism, a virus that hardly meets the criteria for being alive, is billions of times more complex than the most advanced IBM server.

"So!" he announces. "What's the conclusion?"

Theis means this as a rhetorical question, but I have my own answer. *Information* is a word that has as desperate a need for differentiation as *fever* did two centuries ago. Back then, anything that elevated a patient's body temperature was lumped together as a single affliction. It took a century and a half of grindingly difficult medical research to show that fever was generally the symptom of illness, not itself a disease. Any number of pathogens could release pyrolin, a natural molecule that steps up internal heat production. As *fever* in 1803, so *information* and *data* in 2003: The terms aren't subtle enough in their discrimination of cause and effect.

Consider two three-word instructions, each encoding the same quantity of linguistic bytes and semantically identical as imperative noun constructions: 1) Do the laundry, and 2) Get a life. Obviously, the instructions encoded in the second command exist at a far higher and more complex level than the corresponding instructions latent in the first order. So, by inference, are the commands encoded by humanity's three-gigabyte genome. They *must* be; no living thing would exist otherwise. The genome probably specifies the environments under which certain processes may take place—and then lets those enabling environmental conditions contribute their own information to the mix. *Here's what you do,* a gene sequence may say, *but you have to wait until the salt concentration is such-and-such.*

I don't know this for certain; neither does Tom Theis, nor anybody else. But a certain meekness seems to be stealing into the cutting edge of computation technology, even from so an unlikely a source as IBM. It's a tribute to this guy's vast knowledge, competence, and smarts that he dares to be so self-effacing. Only the shysters pretend to omniscience.

Natural computation, Theis admits, does vastly more than its artificial counterpart. Babies self-assemble, self-program, and teach themselves how to breathe, eat, walk, talk, and make up stories. They are genetically pre-programmed to recognize, absorb, and generate nongenetic information. A blank-brained fetus, born a bundle of differentiated cells in sixteenth-century England, in twenty years is Shakespeare. What are all our technical milestones in the face of so monumental a set of natural achievements? How dare we call ourselves inventors at all?

The future of computing, Theis hints strongly, is to depart from where it's been these last five decades. It must escape from its digital prison and compute as nature does: by analog means. Sooner or later, probably sooner, ones and zeroes will give way to computational values that vary smoothly, with steps between defining limits so tiny they may as well not exist. We humans, who have carried the power of natural brains to its greatest known limit, must go back to our own source: nature. And nature, when she computes, does so using naturally evolved analog techniques. Only we humans know from digital; it may well turn out to be a passing fad.

Furthermore, we must visit nature not as conquerors but as acolytes. At last we know enough to be modest; and armed with that new modesty, we must change the way we think, make, and dream. We must do as Lady Nature does.

Does anybody but me see what's going on here? Apparently not. The butt-kickers and butt-kissers still lean forward, waiting for an opening to mock or adore. But what I've just heard seems like Isaiah endorsing Baal, or George W. Bush confessing he's an al Qaeda agent. So, retrospectively, here's my take on the significance of Tom Theis's announcement.

A guy whose firm has for the last two generations been committed heart and soul to digital data processing, has just publicly revealed his despair with Ma Binary. He doesn't believe that digital computation will ever be as good as living systems' analog/parallel computation, at least not at most of the processes that really matter—recognizing faces, creating artwork, and the like. (Heck, I may as well say it: writing.) Simon Haykin came close to this concept at Purdue; now I'm hearing it from one of the highest-ranking technical execs at IBM. *There's no future in digital. It won't wash. Ladies and gentlemen, place your bets . . . elsewhere.*

Shazam!

A little reflection explains why Tom Theis's predicted shift from digital to analog is so important. Historically, digital computing was an anomaly until IBM came along. In both early IT and nature, analog was and is the rule. Analog handles data with material objects or physical quantities, not the rarefied abstraction of digital circuits. A slide rule is an analog computer. So are an abacus, an hourglass, and a sundial. Conversely, the prototypical digital computer, giving us the very term digital, is counting on your fingers. A digit used for counting is either up or down: there's no midway state.

A lot of what we use each day is analog. Radio sets that tune with knobs, dimmer light-switches that turn incandescent filaments up and down, meat thermometers that show twice the temperature with twice the height of mercury column: That's analog. *We* are analog. That's how our brains and bodies work.

The trouble with an analog signal is that it's prone to noise, defined as noncognitive signal components that hide useful information. In AM radio, noise arrives as static. Analog computers have an equivalent gremlin in the form of S/N, or signal-to-noise-ratio.

Static not only afflicts analog systems; it does so in an analog way. The noise begins down at the edge of hearing as a tiny hiss and climbs by slow degrees to a waterlike roar that drowns the signal. While annoying, these effects are intuitively easy for the human brain to understand. Digital systems exhibit far different behavior when afflicted by S/N. Often their responses are counterintuitive. If any of your digital possessions has ever taken an incomprehensible hop, digital S/N is probably the cause. Your computer throws a snit-fit during a routine operation and crashes. Your Walkman skips a track or briefly doubles its volume, turning your eardrums into mush. Hey! That's digital, folks—maddening and miraculous at the same time. Digital S/N glitches are a machine's PMS.

In a sense, the rise of digital computers was technology's admission that it hadn't got its act together. Digital works best under messy conditions. In place of analog's one-to-one correspondence with natural parameters, digital substitutes approximations of varying crudity. Make analog more efficient and digital's advantages start to disappear.

The heart of digital is the 1/0 circuit, or logic gate. In digital computation, a circuit cannot and does not vary in strength with the effect it samples. A digital circuit has only two possible states: on (1) or off (0). The Something/Nothing take of digital is an easier distinction for noisy, high-speed hardware than the far subtler analog scale, which could be summarized as *Zero/ Something/More/Yet-More/Still-More/Lots/Oodles.*

Digital underpins most of our world's newer technology and nearly all of its new economy. Front and center in all this since the creation of modern computing has stood one company: IBM. But it wasn't always like that. Computing, like brain surgery and Dove beauty soap, got its big boost from WWII. Early computers such as UNIVAC and ENIAC grew out of the three main goals of warfare. One was *logistics:* apportioning resources so that Allied armies could get where they needed to be with superior force. Another was *encryption,* decoding enemy messages and encoding ours. And the third main goal was *fire control.*

If you're the gunnery officer of a battleship trying to send 12-inch shells high into the stratosphere and then down to a point 20 miles from your shipboard guns, your computation problems are horrendous. The wind is 15 miles per hour from the east at ground level; atop the shell's trajectory, the jet stream is blowing 220 mph in the opposite direction. In the shell's two-minute flight time, the earth will turn an appreciable amount beneath it, spoiling your aim by 300 yards unless you make corrections. You're dealing with ballistics, which is the science, technology, and art (and probably witchcraft) of directing objects at the outset of unguided flight. That is, your only chance to hit your target is at time of fire; afterward there are no corrections. What do you do? More to the point, how do you do it?

If you were aboard the battlewagon *USS Missouri* in 1944, you would have used on-board analog fire control. In its heyday, this system represented every parameter a gunnery officer had to consider by some strictly mechanical quantity—sliding of verniers, whirring of gears. The result generally fit reality so well that according to direct observers, Mighty Mo could lob a shell into a household garbage can twelve miles away.

As recently as 1968, modern digital systems were nowhere near as accurate. When the U.S. Marines needed an offshore gun platform from which to hit enemy shore targets in North Vietnam, they de-mothballed Mighty Mo. Yet such is the strength of our modern world's romance with digital that these early triumphs of analog have been relegated to the history books.

Interesting as such examples are, they are merely lead-ins to what we might—no, will—accomplish once we apply biomimicry to the analog methods that nature uses to capture, crunch, and transmit data. I've already mentioned a Silicon Valley VC whose client appears to have found a way to stream real-time video over a cell phone. Their "new" method has the elegance of that analog wizard, nature. The scientists inferred from experiment and observation how the human eye-brain system works and reproduced it *in silica.*

(A wordsmith's note. The phrase *in silico* has gained an unfortunate toehold in technical and scientific literature. It is supposed to mean "in silicon"—that is, modeled in a computer—but it is in fact a linguistic barbarism. It is an imitative form, a back-formation from the long-standing phrase *in vitro*—figuratively meaning "cultured in a laboratory," literally "in glass." But *vitrum,* "glass," is a neuter Type III noun whose ablative form has the suffix *–o. Silica,* which is Middle Latin for "sand" or "hard silicon dioxide," is a feminine Type I declension, whose ablative form ends in *–a.* The strict meaning of *in silico* is "inside the pebblestone"—from *silex,* genitive form *silicis,* ablative form *silico.* End of rant; thank you for your attention. The book *qua libra* will now resume.)

The truth was (and is) that analog works. It works in artificial systems and also, spectacularly and unaccountably, in life. Now IBM—the last, best hope of digital—is saying it out loud in public: Analog is the future of computing.

But "wisdom crieth without, she uttereth her voice in the streets, and no man regardeth her." Tom Theis's audience is focused on some nano-scale artifact he's showing them, a fiddly carbon-based component the size of a protein molecule. They don't realize they're really looking at a new and better buggy whip. The whole category this item belongs to is about to go bust. At least that's my take on things.

SCARAMOUCHE REDUX

As the conference winds down, few remaining presentations seem worth attending. There are a few exceptions. Dr. Angela Belcher, who leads a team at the University of Texas, gives us a riveting review of her work in harnessing natural systems to make nanoscale mechanical and IT parts. She's looking toward the day, five to ten years down the pike, when we can give a bacterium top-down instructions to carry out manufacturing operations, or ask a virus to sinter atom-sized electrical leads to a nanoscale transistor, and the little beasties will carry out our will.

But by and large the talks are uninspiring. Something billed as an overview turns out to be the CEO of a struggling pharmaceutical firm who stridently shills his company's wares. It's as if I filled a hall by advertising "A Review of Western Literature" and then spent ninety minutes reciting my own poetry.

Then Scaramouche reappears—Bill Warren of Stillwater, Oklahoma. His slot is billed, none too attractively, as "How Small Objects Can Lead to Environmental and Health-Related Opportunities," but the boring title conceals a splendid talk. Bill starts off just as I'd hoped— with wild humor. Up on a large screen behind him comes a computer-generated video of a fat man blowing up a big rubber boat by lung power. As he's finishing, his fat son bounds down a staircase to the beach and falls on the boat, blowing the man's head off. I'm on the floor at this point; most of the audience is, too. "Play it again!" we call, and he does. It's even funnier the second time. I find it the only instance of intentional humor at the entire conference.

At length, order returns. "I just wanted to show you," Bill grins, "that my firm is developing artificial human organs. That's one of the things we do. We also have projects in f-s lasers." Voice from the floor: *Do what?* "Ultra-short-pulse lasers," Bill says, "f-s standing for femtosecond. A 10 f-s laser burst creates a pulse of light three millimeters long and lasting only 0.00000000000001 seconds. DARPA is funding the work." Bill means the U.S. Defense Advanced Research Projects Agency, his former employer and a current Sciperio client.

"Using these lasers," he says, "we can dissect an individual cell. We can even perform this cell surgery *in vivo* without damaging the tissue adjacent to the cutting area."

Another area that Bill Warren's spin-off firms investigate is nicknamed HAT, for Human Artificial Templating. This new technology is working toward an artificial lymph node. That would supplement diseased or even normal human immune systems, which have difficulty detecting and killing such molecularly clever enemies as malaria trypanosomes and cancer cells.

It's the final bit of technology from Scaramouche PLC that really fascinates me. Sciperio is using nanotechnology to engineer new materials that catalyze the conversion of water vapor to liquid water at room temperature. This might mean a quick end not only to desalination plants, but to the looming global freshwater shortage as well.

"This is an average meeting room we're in," Bill says, sweeping his hand from side to side. "It contains about ten liters of water in the form of vapor. You can extract this vapor fairly easily, by adsorbing it onto a desiccant such as activated carbon. But then you have to rip the water off that desiccant to get a usable liquid. That means you need to refrigerate the desiccant, a process that uses vast amounts of energy.

"We've developed a nanopore form of carbon that adsorbs water vapor, concentrates it, and then lets us extract liquid water from it without refrigeration. It has nano-engineered surfaces that change their nature. They start off hydrophilic, or water-bonding. In this initial state, the surfaces attract and hold water molecules. We can then instantly switch the surfaces so they're hydrophobic, or water-repellent. The water runs off the surfaces and can easily be collected. You're looking at one-fiftieth the power requirement of refrigerative extraction."

The technique works better in high-humidity areas, Bill admits—
"though naturally you want water most where there's the least of it." Yet
even in a desert, the system collects pure water over time. DARPA is funding this project, too; it's easy to see why. Combat soldiers freed from the
need to lug water around would be that much more effective as a strike
force. The logistics are much simpler, particularly for troops operating far
from friendly resupply bases.

This technology also has a flip side. Not only can it extract water without the energy cost of refrigeration, it can provide refrigeration for less
energy than ever before. Bill Warren sees this as a perfect way to keep military supplies, especially human blood, cool in hot surroundings. The
system, Bill tells us, has two and a half times the cooling capacity of water
ice at 0° F.

I see another use: It's a vast potential boost for Sun Belt power grids.
If for half the year air conditioning takes up half the electricity output
south of the Mason-Dixon line, and if that energy drain can be cut by
fifty times . . . .

You complete the equation. My rough calculations suggest we might
be able to kiss the whole Middle East good-bye. Thanks, guys, it's been a
slice. Solve your own damned problems. See you around.

So good luck, Scaramouche, and God bless. There's method in your
madness: A madcap sense of humor seems to be the constant companion
of creative thought. Sometimes you can plod your way to a nanotech
breakthrough, but a lot of the time (*cf.* K.E. Drexler, Ph.D.) you can't.
You've got to make a leap to get there, which requires a kind of brilliant
madness. The ancients used to think a god breathed on you if you
thought this way. They called it inspiration.

SAM JOHNSON, AMERICAN

The jet's front wheels leave the runway; the plane's nose rises, and my seat
tilts back; we're airborne. The big craft climbs, then banks so steeply over
the Golden Gate that the pilot seems to spin around his starboard wing.
San Francisco is right below me, Silicon Valley lost in the haze to the
south. And for some reason, I think of Samuel Johnson.

"When a man is tired of London," Dr. Johnson said in eighteenth-
century England, "he is tired of life." I'd say the same for the United
States of America. The States takes a lot of flak outside its borders: It's too

big, too arrogant, too swaggering—or so they say. I say that's garbage. The United States isn't self-absorbed so much as it's self-sufficient. Its borders—geographical, cultural, technical, ideological—are so vast that the U.S. subsumes a bit, and sometimes a lot, of everything. It's a microcosm of the planet. And more than a microcosm—it's better. The USA is earth's distillate, the best of the best. To say the United States is "concerned with itself" is simply to say it's concerned with everything and fascinated by everything. Right, Your Honor, guilty as charged. Since when are curiosity and achievement crimes?

Come on, you carpers and cavilers. You don't fault Samuel Johnson for favorably comparing his London with all creation. Intellectually and artistically it was—and not only to old Sam. Through his writings, we see his London as he saw it: sprawling, various, inexhaustible. No sense yapping at him from a distance of eight thousand miles and three hundred years. Sure, Sam's viewpoint was insular, but so what? Better an insular Sam than a sophisticated twit who says nothing equally well in several languages. Quality counts more than quantity, right?

Now look at the USA: quality *and* quantity. Not just some of everything: lots and lots of everything. Every race, every language, every religion under the sun. All the scenery, all the trades and professions, all the wild varieties of past and future business. Every problem, sure: hatred, bigotry, shortsightedness, greed, the whole sordid panoply of sin. Scoundrels and wackos and Saturday night specials in slums. But—as Stephen Benét said in *The Devil and Daniel Webster*—"He admitted all the wrong that had ever been done. But he showed how, out of the wrong and the right, the suffering and the starvations, something new had come. And everybody had played a part in it, even the traitors."

You got it, Mr. Benét. So did you, Dr. Johnson. America: the cream of the cream, the nation of creation. God bless.

CHAPTER 5

QUANTUM WEIRDNESS

NOT MY WORRY

NANOSCIENCE and nanotechnology share a dubious distinction. No enterprise on earth, except perhaps professional sports, exhibits less humor that is conscious, or more humor that is both absurd and unconscious. A close study of nanotech can make you feel you're drowning in pomposity. Folks like Scaramouche are few and far between.

This is odd because a sense of humor, at least the type that gets off on true wit rather than pies in the face, correlates well with intelligence. It follows that smarter people, including the more intelligent scientists and engineers, generally have a great sense of humor. Whence, then, the popular belief that technical folk are a dour lot with little juice in them?

Part of the answer may lie in the subject matter. Scientists, like everyone, laugh at things that illuminate the puzzles and paradoxes of their daily lives. Much of this only other scientists will understand; since they share a worldview, they get similar jokes. When you've spent twenty years studying mitochondria, the thought of an imaginary energy currency called systemic adenosine transferase protein, or SATRAP, may leave you on the floor. Not so those outside your world, even that sniffy geneticist in the adjacent office. And as for the layman . . . well. Scientists have no sense of humor, right?

In the early 1980s I wrote speeches for Dr. Larkin Kerwin, president of the National Research Council (NRC) of Canada and a very daunting man. His doctorate was in solid-state physics, taken at the Sorbonne; before this he had been schooled by an elite Jesuit order. He was a small, clean-shaven man with silver hair and owlish horn-rimmed eyeglasses, but he impressed people as if he were John Wayne. I've seen senior Shuttle astronauts smirk and fidget while talking to him, like Welsh coal miners rolling their caps through their hands when Jones the Mine Owner wishes them Happy Christmas.

Before becoming president of NRC, Dr. Kerwin was Rector of Laval University in Quebec. His sense of punctilio was legendary: When seated for a meeting he would stand only if a lady, a member of Parliament, or a Nobel laureate walked into the room. To almost everyone, including me at first, he was icy, forbidding, and completely humorless. A broomstick up his arse, said my boss at the time.

In appearance, Dr. Kerwin and I were utterly unlike. Back then I sported a full beard, moustache, and hair to the shoulders; in the heat of summer I wore African tops, short-shorts the size of postage stamps, and bare feet that gripped a pair of clattering wooden clogs. Jesus of Ottawa, they called me.

By contrast, and however warm the weather, Dr. Kerwin invariably wore full blue suit, white dress shirt, and silk tie. In a long session mapping out the strategy and content of an important speech, he might occasionally shed his jacket, but if so, he immediately hung it in the impeccable closet of his vast corner office. I never once saw him with rolled-up sleeves or a loosened tie, even when he put on a heavy rubber suit and descended into a working coal mine beneath the sea.

But over the months we worked together, Dr. Kerwin began to thaw. Under our two wildly disparate shells, it turned out, he and I were like two peas in a pod. Both of us shared the same passionate excitement about technology and science. The outer reserve of the man was camouflage: It concealed an awe before the natural world as great as Aladdin's in the Cave of Wonders.

Most of the time this childlike freshness remained strictly concealed. Dr. Kerwin's usual manner had the homespun warmth of a municipal report on sidewalk construction. Yet now and then a certain zany wit would surface. I was sitting behind Dr. Kerwin and his senior vice president on a bus one day when the two scientists looked out their window at a gaggle of whores working the streets of down-town Vancouver.

"Is there a collective noun for that trade?" Dr. Kerwin said.

"A flourish of strumpets," the EVP suggested.

Dr. Kerwin nodded. "An anthology of pros."

One day I emerged from a planning session into Dr. Kerwin's antechamber to find the vice presidents assembled for a meeting with him. As we filed past each other, one of the VPs took my arm and put his face close to mine. "What were you *talking* about in there?" he snarled. "That's the first time in my life I've heard the man laugh out loud." I had to think; then it came to me. We'd been discussing speech content, I told the VP. We considered various topics, but Dr. Kerwin wasn't happy with any of them. Finally I'd mentioned bait-and-switch. "Bait-and-switch?" the VP said, uncomprehending. Yes, I told him. I suggested to Dr. Kerwin that we call his talk "Critical Aspects of Governmental Science Policy, 1950–1980." Then when people arrived we'd lock the doors and read them a lecture on Aristotelian epistemology.

The VP goggled at me. "And he laughed at that."

I stared back, deadpan. "You heard him."

The VP walked into Dr. Kerwin's office, muttering. Understand your audience, I always say.

Despite the difficulties of trans-scientific humor, now and then there are breakthroughs. Perhaps a scientist can come up with a jape that the larger community outside her immediate work area understands and relishes. Best of all is a personality like Dr. Bill Warren of Sciperio, whom I call Scaramouche; but such people are rare. Unless they are in positions of authority—and sometimes even then—they are constantly discouraged by colleagues and bosses of lesser intellect who fear their noble discipline is being subjected to adolescent disrepute. Like *I* care, says Scaramouche. Not my worry if you can't take a jape.

Like all clichés, however, the popular view of scientists as humor-challenged has nucleated about a speck of truth. In a working lifetime of talking to technical people, I've consistently confirmed that positive correlation between brains and humor; and science is as full of dullards as writing, plumbing, or anything else. Big Science frowns on such observation: It likes to present itself as uniformly smart. But insiders know. It may be a minor colleague who just doesn't get a new theory and is on his way to a career in real estate. But equally, it may be a dean, lab director, or even a university president who's politically astute but whose lifework at the bench involved little more than reproducing better minds' discoveries.

You can always tell the mediocrities. They sit strait-laced and purse-mouthed and talk down to you in long sentences packed with jargon. This contrasts totally with the scientific cream. To a man—I wish I could say I'd spoken to a woman laureate—the Nobel folk are affable in manner and direct in speech. They're not afraid to use analogies; they're colloquial even when they're precise. They're dreams to interview. It's the fourth-rater who likes to imply he could really communicate with you only if, like him, you spoke middle Assyrian. Dick Feynman, who towered above a run-of-the-mill genius like Freeman Dyson the way Dyson towered above an average physicist, was Scaramouche to the life. Who else would be divorced by his second wife for playing bongos in bed?

MEASURING DWARFS

That being said, there doesn't seem to be a great deal of humor in science these days—at least not beyond the borders of a given discipline. This seems particularly true in nanotechnology. There's a lot of unintentional absurdity, which I'll survey in a moment. But in eighteen months of intensive research, the only funny stuff I found in all of nanotech that knew it was funny turned up on a website posted by Dr. Ossie Tee, a professor of chemistry at Concordia University. This stuff is sly. It starts off fairly straight and gets bent only slowly. I reproduce it here with permission (the URL is http://www.chemistry.mcmaster.ca/csc/orgdiv/nanonano/html).

A Concise Dictionary of Nanoterminology
During the course of our recent research on the chemistry of Cyclodextrins, I have often used the term "Nanobuckets" in the title of talks, in a vain attempt to get onto the "Nanomaterials/Nanotechnology/Nanochemistry" bandwagon. Frequently, I have been asked what the term "Nanobuckets" means and usually I have directed the questioner to the Canajun Dictionary of Unconventional Slang, where the following entries may (or may not) be found.

nano—comb{ining} form = a factor of 10^{-9}; e.g., *nanosecond* [fr. Gk *nanos* = dwarf].

nanoampere—a ridiculous unit of electrical current; of no earthly use, except to physicists.

nanobucket—a molecular-sized vessel, such as a cyclodextrin (q.v.), in which interesting chemistry may sometimes occur.

nanogoat—the female of a small, frisky, short-haired, ovine quadruped (Capra nanoaegagrus), kept mainly for its delicate meat, milk, and cheese.

nanogram—a greeting or message delivered by a dwarf.

nanomancy—the art of divination by talking to leprechauns, elves, or midgets.

nanomania—an excessive enthusiasm for the company of short people.

nanometer—a device for measuring dwarfs.

nanometre—a tiny distance; normally useless, except to spectro-scopists.

nanomole—a nearly invisible, blind, burrowing mammal.

nanomouskouri—a short, bespectacled Greek folk singer.

nanomoussaka—a Greek dish made from aubergines and ground nanogoat meat.

nanophile—a person who is fond of dwarfs (cf. chionileucohep-tananophile—a person in love with Snow White and the Seven Dwarfs).

nanophobia—a hatred or morbid fear of dwarfs. A closely related condition is chionileucoheptananophobia—a dread of Snow White and the Seven Dwarfs.

nanosecond—the attention span of a carrot.

nanovolt—a small unit of potential difference (named for Conte Alessandro Nanovolta, a vertically challenged Italian physicist).

nanowatt—a wee Scottish heating engineer.

AND NOW, OUR FOUNDER

K. Eric Drexler published the first scientific paper on molecular nanotechnology in 1981. In addition, he taught the first course on the subject [at Stanford University] and chaired the first two conferences [on nanotech]. He is currently President of the Foresight Institute [now Chairman of the Board] and a Research Fellow of the Institute for Molecular Manufacturing. He wrote Nanosystems *while a Visiting Scholar at the Stanford University Department of Computer Science and continues to lecture at universities and corporations in the U.S., Europe, and Japan. He received his doctoral degree in molecular nanotechnology from MIT.*

— Jacket biography from *Nanosystems: Molecular Machinery, Manufacturing, and Computation* (New York: John Wiley & Sons, Inc., 1992)

There you have him: K. Eric Drexler. The guy who took up the gauntlet Dick Feynman threw down to the scientific-engineering establishment in 1959. The linguistic genius who coined the word *nanotechnology* as a brilliant parallel of *microtechnology:* the same concept, but a thousand times as small. The lone visionary who risked certain derision from the established authorities and dared to dream. And not only to dream, but to dream the way a reputable engineer does it: not merely in words but in numbers. A quantifying dreamer like Edison or Newton or Feynman himself. A pioneer, a Brigham Young, who surrounded himself with a coterie of the like-minded and welded them into a movement that shook the world.

Dr. Ralph Merkle, for example. Ten years ago Dr. Merkle was a high-profile cryptographer, a member of the research staff in the Computational Nanotechnology Project at PARC—Xerox Corporation's Palo Alto Research Center. Then Eric Drexler moved west to study there. Now Dr. Merkle is vice president of technology assessment at the Foresight Institute, a nonprofit organization formed in 1986 by Drexler and Christine Peterson (a.k.a. Mrs. K. Eric Drexler). At the 2002 Nanotech Planet World Conference in San José, Ralph Merkle gave the second day's keynote address. Two days before that, Merkle's colleague Neil Jacobstein, chairman of the Molecular Manufacturing Institute and an affiliate of the Foresight Institute, gave conference delegates a briefing on what nanotechnology is, where it came from, and where it's going.

In Jacobstein's vision, which is orthodox Drexlerianism, K. Eric is front and center—and necessarily so. Drexler is to the nanoboosters as Christ is to the Christians, as Strom Thurmond is to the southern mossbacks. Most people would agree with this assessment, even when Drexler's views and opinions fill them with skepticism, vague unease, or disagreement amounting to loathing. Love him or hate him, Drexler is impossible to ignore.

Scientific American, doyen of U.S. popular science periodicals and a magazine so reverend that *Esquire* once satirized its motto as "Founded A.D. 11," gave Drexler a two-page inside spread in its September 2001 special issue. That's more than *SciAm* gives to some Nobel laureates. In this article Drexler synopsizes arguments set out at length in his more scholarly books, especially the 556-page tome *Nanosystems.* In *SciAm,* Drexler takes up where Feynman left off. A few excerpts catch the tenor of his *SciAm* message:

"It would be a natural goal to be able to put every atom in a selected place ... with no extra molecules on the loose to jam the works. Such a system would not be a liquid or gas, as no molecules would move randomly, nor would it be a solid, in which molecules are fixed in place." Drexler calls this new, fifth state of existence "machine-phase matter." It will, he says, be characterized by nanoscale machines creating things from the bottom up, atom by atom: new drugs, synthetic materials with vast structural efficiencies, and even copies of themselves—robots building robots building etc.

All this would lead to what you'd have to call an RIE, a Revolution in Everything. Transportation would improve; colonizing near-space and other planetary surfaces would become cost-efficient. Most striking of all, "medical nanorobots . . . could destroy viruses and cancer cells, repair damaged structures, remove accumulated wastes from the brain and bring the body back to a state of youthful health."

Even an RIE is just the beginning. As Drexler sees it, machine-phase nanotech would give us "the eventual ability to repair and revive those few pioneers now in suspended animation (currently regarded as legally deceased), even those who have been preserved using the crude cryogenic storage technology available since the 1960s." [I love the term coined by sci-fi author Larry Niven for these latter-day undead: "corpsicles."]

Concluding his *SciAm* piece, Drexler admits that the Shangri-la he depicts, while on its way, won't happen overnight. He sees "the technology base underpinning such capabilities as perhaps one to three decades off."

Note that the RIE that Drexler prophesies will be neither chemical nor biological, but mechanical. Everything will be accomplished by little machines, much like those we see today whipping about on robotic assembly lines, but smaller—way, way smaller. The Drexlerian "nanobot," he confidently asserts, will be one million to ten million times smaller in diameter than current mechanisms: in volume, one trillion to one quadrillion times more tiny. One quadrillion is the number of quarts of water in Lake Superior.

Nanosystems, the book behind this astounding summary, is not what you'd call a light summer read. Given its mind-numbing opacity, it's hard to see why this book is so effective at raising the hackles of many mainstream scientists. Sample text: "The variations in the potential $V(x)$ associated with sliding a component over a surface can in the standard molecular mechanics approximations be decomposed into a sum of the pairwise nonbonded potentials between the atoms in the object and those in the surface together with terms representing variations in the internal strain energy of the object and the surface." That's not exactly "Arise ye prisoners of starvation" or "We hold these truths to be self-evident," but it's compelling to a certain subset of mechanical engineers. *Nanosystems* has become a classic, at least in Mark Twain's definition: "A book that people praise and don't read." Even people who swear by it, or at it, have rarely gone through this brick of a book cover to cover.

So detailed is Drexler's exegesis that its sheer mass can start to sway you: Your eyes glaze over at the scope of it all. Surely all this amazing technical erudition must lead somewhere? In a kind of *argumentum ad hominum technocratium,* you ask yourself: How could anyone so learned be wrong? "Almost thou persuadest me to be a Drexlerian."

And then the niggling doubts enter your mind. To begin with, making a machine is just the beginning of a workable technology: You must also maintain it. You must control all the separate parts of your machine, mastering matter at scales ten to a thousand times smaller than the scale of your overall invention. To make a building, you must make, lay, and maintain its bricks.

Take a standard industrial robot used on an assembly line, say, 2 × 2 × 1 meters in size. To fabricate this, and then to service it, you will need to control factors all the way from the submicron range (e.g., metallurgical microstructure) up to the half-meter size of the major parts, and everything in between. What happens when a nanobot breaks down? Who squirts oil or its nano-equivalent into lube nipples, or remachines bearings, or sharpens tools? Don't tell me it won't or it can't fail: It must, unless Drexlerianism admits itself to be outright theology.

Consider a soberly presented Drexlerian "invention" such as the Stiff-Arm Nanomanipulator (pp. 398–410). Here in a package only 100 nm high we see a robotic arm with telescopic and rotary joints, core plates, drive gears, snap-on attachments that transport and handle tools, and complex intersegmental bearings. Some of these parts are fifteen angstroms across. Essentially it's a robot from a General Motors assembly line shrunk one hundred million diameters. Beyond that, no concessions are made to the *otherness* of the nanocosm, about which we are only starting to learn—the nano-realm's vast difference from our macroworld.

In essence, Drex does an amazingly thorough job of assuming that the nanocosm will prove to be just like the macrocosm, only smaller. He then expends an equally amazing amount of energy, insight, and erudition expanding on this initial, wrong assumption. On p. 327 of his book, for example, he writes: "An approximation for the pressure gradient along a tube containing a fluid in turbulent flow is the Darcy-Weisbach formula … where v is the mean velocity of the fluid, and f is a friction factor that depends on the Reynolds number R of the flow and the roughness of the wall. The parameter f can be evaluated by methods described in Tapley and Poston (1990); a high value (for a rough pipe at low R) is 0.1, a low value (for a smooth pipe at $R>10^7$) is 0.008."

The exactness of this analysis is impressive. There's only one problem: We don't yet know if it's right. Odds are good, in fact, that it's a half-bubble off level. Corrections such as the Reynolds number are entirely empirical. They were derived, refined, and verified by close observation of how the macroworld behaves. Drex almost never presents truly theoretical explanations for such correction factors. They work, is all—at least at the macroscale. But at the nanoscale, a totally different world presents itself for our understanding. And the nanocosm is not by any means a world we understand. When a pipe is <1 nm in internal diameter, for example—as a single-walled carbon nanotube is—we have no way of knowing if anything other than electrons will flow through it in classical patterns. It's more likely that the well-known stickiness of the tube's carbon atoms will instantly immobilize any material within. We cannot even use macroscale, commonsense terms such as *fluid* at nanometer dimensions. Many organic molecules that constitute fluids in the macroscale would be too big to fit inside a buckytube. If they did squeeze in, they'd quickly jam.

Even molecules that would theoretically fit (diatomic hydrogen, for instance, which measures only a quarter of a nanometer edge to edge) would not necessarily behave in a typically fluidic way inside a nanotube—at least not as a top-notch empirical scientist like Reynolds would have understood the term fluidic. In the nanocosm, a bundle of H_2 molecules would act like a collection of solid, incompressible nodes; they would be no more fluidic than boulders rolling through a trash can.

Even in the macroscale, gases and fluids vastly alter their behavior when key parameters change. To the aircraft designer, the thinner air found in the stratosphere no longer behaves as a classical gas, nor does it exhibit classical fluid-like properties. Aeronautical engineers have learned through hard experience that they must treat ultrahigh-altitude air not as a gas but as a "flux"—that is, a barrage of discrete particles. This numerically measurable change results from change in a single thing: air density.

The moral is clear. You cannot assume that the undiscovered realm of the nanocosm will be just like your kitchen counter. You have to go and find out how things in the nanocosm behave. Otherwise your theorizing is so much fluff.

Drexler, unheeding, proclaims the nanocosm to be SOS, Same Only Smaller. All we have to do is take the machinery we see around us, and shrink it. Itty-bitty bulldozers. Molecule-sized manipulators. Conveyor belts that lug a couple of atoms at a time. Drex even proposes a wholesale return to the Babbage Difference Engine, performing computations at the nanoscale with cams and push rods. Nano-abacus, anyone? There are flywheels for energy buffering, with radii of 195 nm and a rim velocity of 1,000 m/s, which I calculate would give the nano-flywheel a spin rate of fifty million RPM!

No two ways about it: There's no timidity in the man. Here's Martin Amis on Norman Mailer: "He is never afraid to risk looking like a fool. Though perhaps someone should explain to him that there is a role for fear."

No. Something about the whole Drexlerian school sets off alarm bells all over my brain. Drex could be exactly what his acolytes think: one of those crowning visionaries that comes along every century or so, someone who sees so far ahead that when he reports back what he's seen, we groundlings think he's bonkers. The da Vinci of our time. That's what he has a chance of being—maybe. Still, I live by Hemingway's dictum: "The most essential gift for a good writer is a built-in shock-proof shit-detector." And in this case, to use an analogy spoken by the very moderator who introduced Dr. Merkle, the needle on my B.S. meter has swung over so far that it's resting on the pin.

Here's my own take on Drexlerian nanotech. I don't think Drex has established an engineering school at all. Reputable schools undertake original research, then present their findings in peer-reviewed journals and colloquia. They limit their predictions to short extrapolations of current work. These extrapolations must be technically rigorous and mathematically defensible. Sound, solid schools do *not* indulge in wild fantasies and they do *not* put vast effort into persuading the laity via PR. Instead, they try to sway a knowledgeable élite with watertight argument. "The world," Sir Thomas More said at his trial, "must construe according to its wits; this court"—the court of science—"must construe according to the [natural] law."

A casual reading of *Nanosystems* is all it takes to trigger the alarms. This is a utopian vision, and as G. K. Chesterton said of utopias, they first assume that no one will want more than his fair share, and then are ingenious in explaining how that share will be delivered to them via balloon. Utopias, in other words, are skillfully built on iffy foundations. Even when their upper structures are well carpentered, the footings that support them won't pass close inspection. However good a utopia may seem, it's built to impress and not to last.

Drex tries to have it both ways. He wants the unfettered freedom of the blackboard jockey, the absolute liberty of the theoretician, whose mind roams time and space at will. On top of that, he wants the rigor of the experimentalist who presents to us the nature he or she observes and dares us to challenge its reality. Unfortunately, Drex vies for these distinctions without meeting the standards and qualifications of either. *Nanosystems*, and the doctoral studies on which it's based, constitute a kind of spadework. With them, Drex shows the engineering world the stick-to-itiveness and the "fine ratiocinative meditativeness" that are the engineer's equivalent of piety. He demonstrates his willingness to boldly go where no man has gone before—including skating on ice so thin it's imaginary. He's an experimentalist who hasn't done and won't do experiments, a theoretician brilliantly connecting data that have never been derived. Drex says he's invented an entire new field, or rather a field full of fields. He calls it "theoretical applied science." Yet Drexler isn't summarizing what he, or anyone else, has ever seen. He's what-iffing, on grounds as nebulous as common sense—what *should* exist at the nanoscale, once we look. There's a word for what Walt Disney called the "plausible impossible." It's *specious*.

I prefer the more learned and modest approach of one A. Einstein, a more substantial scientist than K. Eric Drexler. Einstein defined common sense as the set of prejudices that we accrete before age eighteen. He cautioned against using this as a guide in the extreme worlds we enter when, through laboratory experiment or else the thought experiments of imagination, we accelerate up toward lightspeed or shrink down to the nanocosm. Common sense, it turns out, is neither common nor sensible outside the macroworld in which it was learned—and in which, and *only* in which, it applies. Yet here's Drex, Our Founder, blithely traipsing through a world of quantum weirdness as if it's a backyard barbecue in Redwood City. Mechanosynthesis of diamondoid! Nanoscale symmetrical-sleeve bearings! Nuts and screws and rods in sleeves! Gears and rollers, belts and cams! Dampers, detents, clutches, and ratchets! Drives, fluids, seals, pumps, and cooling systems! Electrostatic motors—and here I have to say, as the musical comedienne Anna Russell says about the libretti of Richard Wagner: "I'm not making this up, you know."

The unease you feel when you consider the technical gaps in Drexler's intellectual fabric increases a hundredfold when you examine the means that he and his followers have chosen to propagate their beliefs. Consider the rhetorical technique called poisoning the well. In a single paragraph called "Criticism of Criticism" (p. xviii of the introduction to *Nanosystems*), Drexler airily dismisses all possible objections to his ideological constructs out of hand. He states that his approach of molecular manufacturing will work if he's given inflexible molecules, nonreactive atoms, stable fragments, and a total absence of any trace contaminants. But in so doing, Drexler immediately paints himself into a corner. His own constraint-set obviates his predicted inventions, such as the nanoassembler or the blood-cruising medical nanobot that repairs cells as they age. (We'd love to help, sir, but first your liver must be made from diamond.) His nanobots will sense their surroundings but not see them; they will not be subject to dislocation by thermal motion; their work will produce no excess heat, etc.

Nanosystems is, in the kindest view, a thought experiment carried far too far. At first it seems the product of a bright, unbridled mind, which like Stephen Leacock's famous general mounts a horse and gallops madly off in all directions. But is Drex really that sophomoric or self-deluded? On deeper consideration I'd say no. His real goal, I think, was to establish himself as *the* expert in a new high-profile area, without the bother of original experiment or rigorous, cautious, defensible interpretation. And this he has most certainly done. He's certainly enlisted a mass of followers. To a certain stripe of pop-sci, New Age nanobooster, the kind that entertainingly infests websites like *planet.hawaii.com,* Drex stands serene and solitary as Our Founder. Even reputable scientists must acknowledge this achievement, though they cannot quite fathom how or why it occurred.

THE CHURCH OF ST. DREX

Let's consider some social effects of the movement called Drexlerianism, starting with the historical context.

Some of the biggest organizations in history owe their success less to merit than to rhetoric. This is as true today, in the third millennium A.D., as it was in the third millennium B.C. Orthodox religions; the personality cults of Mao, Stalin, or Mobutu Sese Seko; computer operating systems with spiffy graphic user interfaces based on ancient, creaky op systems— none of these rest on the objective supremacy of their moral or technical values. Instead, they all constitute triumphs of marketing.

Archaeology has unearthed evidence of vigorous eclectic belief systems operating in Judea circa 100 A.D. These combined elements from Persian astrology and Greek mystery sects, as well as Judaic tradition. In the last twenty centuries, systematic orthodoxy has scrubbed these vigorous, eclectic movements from the face of the earth.

Among the truly learned, the tenets of a philosophy are subject to dissention, amendment, and debate. But in the modern cult of personality, these checks and balances are subsumed into the public person of an all-wise ruler. His actions may be praised, but as holy mysteries they are totally inarguable. What The Man says, goes.

In our own age, the variations on Microsoft's Windows, though nice to look at, have tended to work through a cobbled-together ad hoc code called "disk operating system." DOS routinely gums, jams, and slows down the most modern processing hardware because it has the architectural elegance of a Calcutta shantytown. Before NT at least, Windows was a window on nothing so much as the Eocene Age of the personal computer.

Yet where, in every case, is the competition? The Jerusalem variants of Judaism are literally history. Dictators of all stripes still strut and fret their hour upon the stage. Linux languishes. Many of these improbable victories are due to rhetoric, the art of persuasion by emotion rather than by reason; this effect is truer than ever today. Jung was wrong. It's not dreams that are the royal road to the unconscious; it's the glands. Words that appeal to the glands harden nebulous images and crystallize vague biases, and those who shape these words shape us. It's what I see happening with Eric and the Nanoboosters. Here are the signs by which ye may know them:

1. *Do not explain or otherwise use the technical work as a detailed basis of argument. Refer to it as a body.* It is sufficient for Drexler to have written *Nanosystems.* That being done, there is no need for Drexlerians to review the data or the reasoning the book contains to answer objections. There is no need for additional counterarguments. The book is presented as definitive, authoritative, and fixed: in other words, as Holy Writ. On p. xvii, *Nanosystems* blithely dismisses all possible criticism in fifteen lines, which being interpreted is: *Says here, brothers and sisters!* Abruptly, the argument from data becomes the argument from authority.

2. *Give audiences rhetoric rather than close verbal reasoning or mathematics.* This leads to some questionable techniques, which Drexlerians have polished to a high degree. In his keynote at San José, Ralph Merkle—as good a speaker as I have ever heard: relaxed, confident, droll, natural, and silk-smooth—did a straw poll of his audience. After reviewing the Gospel According to St. Drex, he asked us: "How many believe this will ever happen?" Some hands appeared. "And how many believe it will *never* happen?" Other hands shot up, including mine. "This is typical," Merkle said with a broad smile. "Strong support, with a little skepticism." In fact, neither he nor anyone had counted hands; my quick impression was of an approximately equal number of votes. But by this time the Apostle Ralph was on to his next paragraph, and the pitch rolled on—rhetorically brilliant, intellectually dishonest. Merkle's premise: *Democracy can establish natural law*—a majority believes it, so it's true. (I'm sorry, sir, you can't use the diving board. The plenary session voted to adjust the gravitational constant. If you tried to dive, you'd simply float. Tomorrow, perhaps.)

3. *Make predictions positive, not negative.* Whatever you think of their ideology, the Foresight Institute, the Institute for Molecular Manufacturing, and all their passive-aggressive fellow travelers have got one thing right. They're predicting that certain things *can* be done, not that they *can't* be. That forces opponents to argue such things are impossible, and that's a bad position to be in. Rhetorically, the strategy is inspired. Say that something will happen and nobody can ever prove you wrong. If time goes by and it doesn't occur, you smile and say, "Wait a bit." You can never be proven wrong: that would take infinite time. But if you predict impossibility, you paint yourself into a corner. At any instant you may be proven wrong and look like a fool. This may be why, as one futurist pointed out to me years ago, people forget or forgive all untrue positive predictions, yet remember each untrue negative prediction to the nethermost detail. Say "It will be!" and you're safe. Say "It can never be!" and you may be on the way to being a pariah. The Drexlerians have been undeniably clever in seizing this rhetorical high ground.

4. *Reek with modesty.* Correctly accused of presenting no hard evidence to back up their recent claims of human cloning, the fringe group called the Raelians did not respond with lawsuits—or with the missing data, either. Their spokespeople merely looked serene and said, "We know the truth." Similarly, the Drexlerians don't beat their collective breast or call out the lawyers when their facts or ideas are attacked. Instead, Drex and the gang sail along emanating a believer's utter certainty. Their certainty, however, is not of factual truth but something nebulous, emotional, and rhetorical. Call it righteousness, the kiss of God, the stamp-of-approval of the universe, the life force, history, scientific inevitability—any absolute authority that you fancy. No rational, carping challenger can share that bliss. "Mock if you will! By doing so you merely exclude yourself from the Kingdom of Heaven and destine yourself for hell: We have the last laugh. Poor you. Not that we Drexlerians hold a grudge. Oh, no. You can always renounce your wicked ways and come over to our side. Come to the nanomanipulator's everlasting arms."

One who wants to command respect must do so by deeds and knowledge. Yet nowhere in my hearing did Merkle revert to text and rehearse any of those brilliant arguments made in *Nanosystems*. From first to last, he pitched to the glands and not the brains. It wasn't a poster, nor a paper, nor a presentation: It was a sermon, stage-managed for maximum rhetorical effect. What I witnessed in San José was a rather scary case of homiletics.

Why do Drexler and his followers carry their arguments to the public far more than to the scientific community? Why is their chosen venue not the journal but the pulpit? Several answers might apply.

First, the Drexlerians know how harsh a full peer review would be. Revive a forty-year-old frozen corpse whose cell walls, all six trillion of them, are ruptured by internal ice crystals? Right. Send nanobots to scour away atherosclerotic plaque when legitimate science is just starting to understand it's not an inert deposit but an inflammatory eruption of staggering complexity? You might as well heal a wound by taking a rasp file to the scab. Take it from another professional rhetorician, ladies and gentlemen: This ain't science. It's sci-fi.

When you undertake experimental science, you begin with some idea of where you're going. You have, in other words, preconceptions. But at the same time, you remain open to being convinced and instructed by the nature you're investigating. You accept that your initial goals and premises are just a starting point and may prove to be a crock. Along the way you learn, change, adapt. But when you're writing sci-fi, you settle on where you want to go and stick to it regardless. In science, facts and findings rule all: There's no appeal from nature. In sci-fi, facts and findings are selectively stressed or completely disregarded to prove an inflexible point. Sci or sci-fi: I leave it to you to decide which category Drexlerianism best fits.

Second, while Drexlerians are quick to co-opt others' research to shore up their contentions, they don't undertake original research of their own. Neil Jacobstein made extensive reference to the work of Dr. Wilson Ho, "a nanoscientist at Cornell," as supporting Drexlerian theory. But Dr. C. M. Drain, a chemist from New York City University sitting beside me as this was being said, snorted in derision. He and Ho knew each other well. Dr. Drain said, "Wilson [Ho] doesn't know this guy is talking about his work. Besides, he's not even at Cornell anymore."

While the Drexlerians cite others' work selectively, they don't do so sparingly. In fairness, everyone trumpets supporters and ignores detractors. It's part of the common human talent for denial. But the Church of St. Drex raises this rhetorical one-sidedness to high art. Neither Jacobstein nor Merkle, nor Drexler in his *Scientific American* piece, ever mentioned a critic by name. Instead, they depersonalize: "one prominent chemist" or "another well-known chemist." Vague, nameless detractors, you see, hardly worthy of mention. We refer to them only because of our surpassing humility and thirst for objective truth.

One of these nameless detractors writes overleaf from Drexler in the same *SciAm* issue, and *prominent* is too mild a term for the man. He's Dr. Richard Smalley, recipient of the 1996 Nobel Prize in Chemistry for his discovery of the carbon "nanotropes" called fullerenes. Patiently, surgically, with trademark good humor, Smalley vivisects the whole idea of the molecular assembler. Making one ounce of something this way, he notes, would require moving at least 600,000,000,000,000,000,000,000 atoms, making and breaking a minimum of one atomic bond for each. "At the frenzied rate of 10^9 [one billion atoms] per second it would take this nanobot . . . 19 million years" to assemble a single material ounce, Smalley writes. Besides, "in an ordinary chemical reaction five to fifteen atoms near the reaction site engage in an intricate three-dimensional waltz that is carried out in a cramped region of space measuring no more than a nanometer on each side . . . . There just isn't enough room in the nanometer-size [chemical-] reaction region to accommodate all the fingers of all the manipulators necessary to have complete control of the chemistry." Concludes Smalley, with a wit more devastating by being understated: "Feynman memorably noted, 'There's room at the bottom.' But there's not *that* much room."

Drexler has obviously heard this cavil many times before, and he predismisses it in his own *SciAm* piece. "These are reasonable questions that can be answered only by describing designs and calculations too bulky to fit in this essay." [*It's in The Book!*] Then, in one of those sneers that only scientists and poets have mastered: "These examples point to the difficulty of finding appropriate critiques of nanotechnology designs. Many researchers whose work seems relevant are actually the wrong experts— they are excellent in their discipline but have little expertise in systems engineering." Translation: *You guys wouldn't hate me if you were only smart enough to understand me.*

To those readers who feel my critique is excessive, I offer a koan. Where in all reputable science or technology is there another case like the

Drexlerians? Which genuine discipline polls its audience to arrive at scientific truth, or shills autographed books from its leading lights? One
must distinguish between a *field*—an area of rational enquiry—and a
movement: an emotional state that uses whatever methods it can to but-
tress its *a priori* beliefs. A field has members, convinced by facts and
steadily adding to those facts. It is constantly rational and skeptical. A
movement has adherents, convincing one another by groupthink that
what would be really cool is really possible. As such, a movement is
breathtakingly gullible.

Like all congregations, the Church of St. Drex deals in an indefinite
future, not the present. Its goal is not the possible, but the forever
unattainable. "Work and pray, live on hay; / there's pie in the sky when
you die." The datum that reveals Drexlerianism as a partly natural and a
partly revealed religion, the dead giveaway if you'll forgive the pun, is the
promise of immortality. It's physical! It's mediated by nanobots! And it's
coming, brothers and sisters! It's just not *here* quite yet. Stay tuned.

From a Foresight Institute brochure:

> Participate in the Foresight Institute as a Senior Associate! Sign up
> NOW. **Associate Level,** $250/year. Includes a copy of *Engines of
> Creation,* Eric K. Drexler's groundbreaking work. **Fellow,** $500/year.
> Includes one autographed book: *Nanosystems or Engines of Creation.*
> **Colleague,** $1,000. Includes framed artwork, autographed book,
> PLUS 30-minute conversation with Ralph Merkle or Christine
> Peterson. **Friend/Corporate,** $5,000/year. Includes engraved
> "Friend of Foresight" award, signed artwork and book PLUS 30-
> minute conversation with Foresight Chairman, K. Eric Drexler.

A small note at the end of this list reveals the stated sums are only a
fifth of what you'll pay: Pledges must be renewed annually for five years.
At the end of that time, one supposes, you'll be given an amulet blessed
by St. Drex himself. Act at once! [Git on a wagon goin' West / Out to the
great unknown / Git on a wagon rollin' West / Or you'll be left alone!]
Show this unwashed author, *por favor,* a single legitimate discipline that
engages in such tactics. Even the alumni campaigns of the big universities check themselves before they reach this stratum of vulgarity.

Meanwhile, back at San José, Merkle's performance is masterful; I
doff my hat to so skilled an orator. Merkle adduces and cites reputable

scientists—even, wonderful to say, Dr. Richard Smalley. (Don't knock it. If the devil can quote scripture for his own purpose, the elect can quote the devil to his own discommodation.) IBM's Millipede nanoscriber, UCLA's Dr. Montemagno and his molecular motors, Dr. Nantero's electronic inverters made out of buckytubes—all fly by at breathtaking speed. How do the scientists so cited feel about lending the glamour of their name, the blood of their work, to C-St-D? Merkle never says. Drexler makes his appearance (oh, clever) five minutes into Merkle's address, his name slipped into a long list of experimental scientists (which he most certainly is not). Then all the tenets of Drex-doctrine come out one by one. We'll have artificial blood cells, nanospheres of oxygen compressed to 1,000 atmospheres of pressure: seven tons per square inch. These will float benignly in the bloodstream until needed, then release their precious load at a perfectly regulated, predetermined rate. Goodness, my heart's stopped! Better call the doctor. Hello, doctor? My heart's stopped, what do you advise? Emerg ward within the next half-hour? Right then, see you there. Bye! Have a nice day! Merkle mimes the whole conversation.

Even the corpsicles get into the act. "Which would you rather be part of," Merkle demands rhetorically, "the *experimental* group? Or the *control* group?" The control group, of course, is mortal men doomed to die. Are you saying you want to die? No? Then, laydeeeeeez 'n' gemmun, git onto that train to the Promised Land! All '*booooooooooooord*!

By this time I'm shaking my head in a kind of appalled admiration. If my Hemingway gland is telling me the truth, then this entire performance, including its truth-by-vote component, is not science but marketing. It gives snake oil a bad name.

All this time, slides projected from Merkle's notebook computer have been flashing onto a screen, then disappearing, with blinding speed. The more complex the graphs and equations, the shorter the dwell time. "Simple expressions," Merkle says dismissively, as a set of six simultaneous differential equations vanishes in half a second. Finally the charade ends and the moderator calls for questions. My hand is in the air. Merkle gives me a magisterial nod. Approach, O student of Truth, and ask!

"It seems to me you've ignored the effects of Brownian thermal motion. You postulate the ultimate development, at some unspecified time in the future, of a molecular manipulator of nanoscale proportions. Yet in aqueous solution at standard temperature and pressure, the average molecule vibrates at about ten gigahertz. That is to say, it displaces laterally by up to half its diameter at a constant rate of ten billion times a second. Do you seriously propose to perform detailed cabinetwork on such a thing?"

Bingo. For two and a half seconds, Merkle's face freezes. Then he recovers and attempts an answer. We will, he says, operate our nanomanipulators only on *stiff* substances. Our preferred substrate is diamond, a well-characterized carbon allotrope with a perfectly cubic crystal. It is both hard and stiff. Stiffness is paramount. All we have to do is clamp a diamondoid nanomanipulator onto a diamondoid workpiece and the zero relative motion between the two will permit...

Merkle drones on. A tall, bearded man beside me leans over and whispers a comment: "He hasn't answered your question."

I shake my head. "Ten gigahertz. It's hard enough to kiss your partner on the dance floor."

"Good question, though. Why did you ask it?"

I shrug. "This whole Drexlerian thing is too crude. You can't shrink iron-age mechanical contrivances to the nanoscale and call it quits. It's too inelegant, it shows contempt for nature's subtlety. The Nano-Peeler, Peels Individual Carbon Atoms Quickly and Surely, A Must for Every Modern Housewife Who Wants to Be a Scientist and Prepare Nourishing Mechanisms for Her Family and Graduate Students at the Same Time." I say a rude word.

"A lot of people are sinking good money into it," my friend says.

"A lot of people are going to get hosed." I stick my hand out. "Bill Atkinson. I'm writing a book on nanotech."

"Tom Theis. IBM."

IN ALL THIS LAND of Drexlerian make-believe, there are hopeful signs. Surprisingly often for a lady usually drawn as being red in tooth and claw, nature can be merciful. For example, the inability of HIV to saturate humankind may be due to the Black Death. That ancient scourge selected for people with highly efficient immune systems; they survived and bequeathed their descendants—us—their immunity. Over the centuries, other plagues have shown similar attenuation. When it first hit Europe, syphilis acted like Ebola virus. It rotted brains and sliced away faces, and killed within a few years—sometimes within months. Four centuries later when drugs were available to treat it, syphilis had grown gentler. It was still an ultimate death sentence, but a long-term, manageable one.

From the pathogen's viewpoint, this gentling process is one of enlightened self-interest. When a disease propagates by direct contact, it spreads deeper through a population if it goes more slowly and pulls its punches. This is because a carrier infects more victims if he doesn't die too soon. That's a strategy that maximizes the ineradicability of an infectious agent. Even for a syphilitic spirochete or a malaria trypanosome, bad guys finish last.

I see an identical thing starting to occur with Drexlerianism. By now, Drex has milked his wilder prognostications dry. Twenty years of standing on a corner soapbox shouting "The Kingdom of Heaven is at hand," has been successful to the point of embarrassment. All those fervent converts, all those believers in imminent material salvation, now demand (ever so reverently) that Drex, Merk, and the boys put up or shut up. The result is a Texas-based company called Zyvex, incorporated in 2000 to realize the Drexlerian vision of fast, workable molecular mechanosynthesis *in vacuo.*

At this point, the germs of Drexler's ideas are suddenly encountering nature's immune system—the adaptive response by which she chews up hubristic ideas. Now nature gets to comment on the things that Drexler *et al.* so casually proposed for her. The result? Already there are noises, immensely amusing to us skeptics and debunkers, of theoreticians' faces meeting the brick wall of reality at 150 mph.

Here's analyst Gary Stix's summary, in *Scientific American:* "Zyvex, a company started by a software magnate enticed by Drexlerian nanotechnology, has recognized how difficult it will be to create robots at the nanometer scale; the company is now dabbling with much larger micromechanical elements, which Drexler has disparaged in his books."

SciAm staff editor Steven Ashley is even more wry. "Perhaps Zyvex's trek toward molecular nanotech," he says, "could be financed by small contributions from its legions of true believers."

Brothers and sisters, it's collection time.

CHAPTER 6

SEEING THINGS

LUNCH, WITH TIME WARP

MOST PROFESSIONS pay better than science writing—investment banking, say, or flipping hamburgers—but none I know of has the same job satisfaction. A good interview, one that stretches my *dura mater* and pours in new ideas I'd never dreamed of, leaves me giddy. I walk from the laboratory with my eyes glazed and my shoes two inches off the floor.

One Friday in April 2002, I'd just completed one such interview at a research university. My contact had reviewed some recent work that began with equation-fiddling and ended with what might yet prove to be a workable nanoscale transistor. I floated to the faculty club, where two old friends awaited me for lunch. Then I got smacked a second time.

There they were, looking older than when I'd last seen them years ago: Falstaff and Bardolph to the life. But that's not where the shock lay. Between them, unexpectedly, sat another long-time friend of mine, a woman whom I'd first met in 1968. The men had changed—and she hadn't. She looked exactly the way she had thirty-four years ago, when I'd introduced her to my friend Jack Falstaff and they'd fallen in love.

"You remember my daughter," Jack was saying. I smiled and nodded— wonderful thing the mind, how it runs all by itself in times of shock—and shook her hand. Haven't seen you in years, I told her. You look just like your mother. And so she did. Face, form, height, hair, build, all identical. *Identical.* Same walk. Same manners. Same cinnamon eyes. Mixed emotions? I didn't know whether to spit or go blind. Looking back now I'd call it a rueful, grateful amazement at the persistence of our species. For me, this young woman—no, *lady;* that all-but-vanished mix of excellence, brains, and grace— symbolized humanity's everlasting power to refresh, renew, and prevail.

I'm raving, I know, but it was a trauma. It was as if her mother had stepped out of a time warp. For a couple of seconds I was seeing things.

It was no surprise to me when "seeing things" became our topic of discussion at lunch. Susan Falstaff was close to completing her doctorate in molecular biology. When she heard I was researching a book on nanotech, she lit with interest.

"It's an exciting time," she said. "What's happening now is transforming my discipline. Ever since chemistry began we've taken it on faith that molecules exist and that they're doing what we think they are. But all the evidence has been indirect. We saw the footprints and made deductions. Now we have the instruments to image what's really happening, sometimes as it happens. It's incredible."

"For now we see through a glass, darkly—" her father quoted.

"—but then face to face," I said.

She hardly heard us. "Of course the evidence was good and the deductions were sound. Right from Dalton on—"

"Democritus of Athens, dear," her father said. Jack is a retired high-school history teacher. "Fifth century B.C. *A-tomos* is Greek for 'uncuttable.'"

"Oh, *Dad.* Anyway, chemists knew all those things, the material forms and their processes had to be in there, otherwise we wouldn't find the clues we did. But they were still only clues. Now we can image molecules and even atoms. For my field, for all of science, it's become an absolutely amazing age."

As it turned out, the lady was the first of several dozen other scientists and CEOs I interviewed who said the same thing. Furthermore, in choosing that particular verse from a Pauline epistle, Jack Falstaff showed that even though he was no scientist, he grasped his daughter's metaphor and thus the reasons for her enthusiasm. Various religions preach that a lifetime of trust in an unseen truth will one day be rewarded by an eternity of direct witness. That precise prediction describes science and technology today: They exist in a state of grace. Their faith in their model of the invisible world has been triumphantly borne out; the words of Dalton and his fellow prophets have come to pass.

To pass, and then some. What the new atomic-resolution instruments are revealing does more than confirm the long-held views: It embroiders and expands them until in some cases they are hardly recognizable. The world that scientists such as Susan Falstaff now see face-to-face is far more strange than she or anyone could have imagined. *The more thou searchest the more thou shalt marvel.*

A word of background. "Seeing is believing" runs the adage, and in most cases that's true. *Before my God, I might not this believe / Without the sensible and true avouch / Of mine own eyes,* says Horatio when he sees a ghost. But in the four centuries since *Hamlet* was written, neuroscience has shown us that witnessing is not a direct and simple thing at all—it is a fiendishly complex act.

This is partly because it makes little sense to speak of "the eye" or "the brain" as if each were an independent entity. The term "eye-brain system" cuts closer to the truth. The eyeball is no passive sensor. It contains embedded software that preprocesses the data brought to it by photonic wavefronts. It receives these waves along a narrow sliver of the electromagnetic spectrum, refracts them by transmission through clear colloidal suspensions and gels, brings them to a color-corrected focus on an organic substrate packed with sensors that are specialized for brightness or hue, preprocesses this gigabaud data flow with elegant algorithms we have just begun to understand, and dispatches it back to the visual cortex for final processing.

The optic nerve was once thought of as a dumb connection, a simple ionic pipe that carried raw sensory data to the brain. Now we know that the optic nerve massages its information in transit. It's really a cerebral pseudopod. The brain isn't content to sit and wait for information to sift in. Instead, it lunges out through its light receptors, confronting the world and seizing its data. Nor are the eyes mere windows of the soul, reflecting what lies within. They are the mind's exits, the hatches by which it leaps out into the world. An ancient belief held that the eyes send out light as well as gather it. In a conceptual sense, they do; but the things they emit are ideas rather than photons. Humans continually project concepts onto the world they see, to determine how well these preconceptions fit. Incoming facts are constantly tested against what's known from other sources—both immediate (Is it night?) and retrieved from memory (Have I seen anything like this before?). And because of this natural complexity, there's literally more to vision than meets the eye. In a sense, belief comes before sight. Nobels by the bucketful lie right beneath our noses, and always have. We merely lack the concepts that would let us recognize the data. Now and then, scientists find what they're not looking for. But they never, ever find anything for which they have no previous category.

Most people concede that science must discover, test, and verify its facts before these can be harnessed into new technology. What is not generally known is that *even within the basic science,* theoretical work must be done as a continual, ongoing activity—not only following a discovery, but throughout the entire process that leads up to it. You don't have technology-ready knowledge, in fact, you don't have knowledge at all, when you have no concept what in heck (or in the nanocosm) you're looking at. Every so often, even the greatest scientist must pause and let his brain catch up with his observations. For the conceptual thinker, the defining term is *Aha!* For the experimentalist, it's *What the hell?*

Nanoscience today is dominated by observation and will remain so for at least another five years. This is typical in emerging fields: Every hour produces new data that theoreticians struggle to make sense of. Nor is this struggle limited to those pure thinkers who sit in splendid isolation far from any lab. The experimentalists themselves are combatants in this conceptual arena. To go on observing, they must continually develop new beliefs and refine or discard old suppositions.

Nanoscience exists because imaging instruments got so good, so fast, so recently. That's why there's a theoretical backlog, with mounting piles of data from the experimentalists that beg to be explained by experimentalist and theorist alike. As explanations emerge, so will workable technologies. Nanoscientists themselves have barely begun to see what the nanocosm holds; the rest of us can only wait and wonder. We can guess, but we don't know. To pretend otherwise is Drexlerian hubris.

THE THEORISTS' COURT

In all technology and science, two themes are inextricably tied together. You could call them observation/hypothesis, experiment/theory, or instrument/idea. The key cycle at the heart of science is the endless tennis game played by these two things.

However you picture it, the flashy bells and whistles in a lab are not science per se. They are experimental technology: the material means used to test scientific concepts. Science itself is pure idea. Among some scientists, perhaps even the majority, this simple fact is overlain with the prejudice that technology—indeed, all experimental fact-finding—is somehow inferior to theorizing. True, theory can be elegant. It often requires nothing more than brains, free time, and a blackboard. But theory divorced from the constant corroboration of experiment, while it may be elegantly self-consistent, quickly gets absurd. It is specious rather than accurate; it sounds good without being so. Theory alone never works. It needs experiment to keep it honest. Without experimental proof, theory eventually drifts until it is elegant, self-consistent—and wrong. That's the origin of the orbital epicycles devised to explain planetary motion, or the massless matter called ether that as recently as 1880 was thought to fill all space. For that matter, consider Eric Drexler. Wide knowledge, profound thought, sound reasoning, and immense detail—almost none of which has been verified at the nanoscale by anyone, anywhere, ever. Every sentence in his book *Nanosystems* should begin: "Wouldn't it be nice if . . ." Or, putting a sharper point on it: *K. Eric Drex, K. Eric Drex / The man who dispensed with reality checks.*

Just as a brick wall cannot exist without bricks, so there can be no hypothesis without proven, pre-existent facts. The most profound theory (such as special relativity or the double helix of DNA) exists because it orders facts. If either of the twin siblings of fact and idea has a higher reality or independent existence, it's the data. Even without hypotheses, facts would still exist—if only as a pile of bricks waiting for a master-architect to make them into a wall. But if experiment and hypothesis are kept equal, if neither tries to lord it over the other, they can function as twin engines to good science, dependable technology, and a healthy economy.

The thought/discovery cycle is never complete, nor can it be. Like all tennis games, it needs two players. Each theoretical explanation is an interim stage, a provisional thing. It must itself be explained at the next deeper level of understanding. So delve down, experimenters: Find the struts below the stage. Present them to the theorists, so they may find explanations of their earlier explanations. Then, both of you, start again.

Just as there is no limit to how much scientists want to know, so there is no limit to how closely they want to look. Their motto is an old one: *Multum in Parva* ("much in little"). Technologists invent an instrument that directly images molecules. Theorists explain the natural processes so revealed, then forecast what hasn't yet been seen. Back to the technologists, who test the new ideas and see if what's been predicted is really there.

The experimentalists have already devised instruments that see atoms by sensing their orbitals, elegantly shaped electron clouds that clothe an atom's nucleus. By 2020, we may image the nucleus, the infinitesimally tiny kernel that holds almost all an atom's weight. Again, by indirect means we know the nucleus is in there; it has to be. That's what deflects other invisible particles that we throw at it. But the nucleus is squashed into a volume as relatively tiny as a pea in a planetarium. The day that even a big, bloated nucleus like uranium-235 is directly imaged, however fuzzily, I'll start a sequel to this book called *Femtocosm*.

FACE TO FACE

I entered grade six in the fall of 1956 determined to be a scientist. The trouble was the embarrassment of riches: Where to go, what to choose? Every discipline was burgeoning, both in new facts and in the new theories that construed them. In the United States, Projects Explorer and Vanguard planned to launch an artificial earth satellite for the International Geophysical "Year," an 18-month period starting January 1, 1958. The Avro Arrow, a fighter so advanced that it would still rank among the world's top warplanes, was about to be fitted with the Iroquois high-bypass turbojet. And in materials science, metallurgists announced the world's first photographs of atoms. These had been obtained using the new techniques of high-resolution X-ray crystallography.

Bliss it was in that dawn to be alive; but to be young was very Heaven. I mean: *atoms!* We're blasé about this now, as we are about so many miracles, but back then it was like getting a signed postcard from God Almighty. The so-called atomic photos were specious, as it turned out; the crystallographers had no true images. They'd merely made maps, patterns that showed where atoms lay within certain simple, unflawed crystals. The dots were no more pictures of atoms than a speck on a map was Chattanooga. But while the claims of direct atomic imaging were hubris, they laid out clearly what science hungered to see and devoutly believed it soon would. Collectively, the technical culture had a shrewd intuition where it was to go. It was not *zeitgeist* (the spirit of the age) so much as *futurgeist* (the spirit of tomorrow). That sure and certain hope shone from Dick Feynman's essay *Room at the Bottom*, which launched the conceptual stage of nanoscience in 1959.

Absolutes were on my mind as I progressed through school. They were on everyone's mind. Records fell weekly. Canada's HARP, or High-Altitude Research Project, used a giant cannon to throw shock-resistant instrumentation out of the atmosphere and into near-space, 60 miles above the earth. The U.S. Farside program lifted multi-stage, solid-fuel rockets to 20 miles above the earth via helium-inflated balloon. The rockets ignited, punched through their carrier, and reached for the far side of the moon.

While I was dazzled by these bold attempts and practical achievements, I retained a theoretician's outlook. By the time I reached high school, I began to be troubled by other absolutes: not engineering records, but intangible ideas. Something that particularly gnawed at me was the concept of the edge.

Here's what kept me up nights, if you can believe it. Fundamentally, a knife cuts by pressure, defined as force divided by bearing area. You can increase cutting pressure by increasing force—that's why you bear down harder on a dull knife—or by decreasing bearing area, which you do when you strop up a knife's edge. A keen blade cuts with so little effort because the force you do apply bears on so small an area.

All good, sound materials science. My problem was this: From my studies in Euclidean geometry I knew that an angle was an ultimate—an absolute, a Platonic ideal. I didn't know the word *fractal* in 1964—I'm not even sure that Dr. Benoit Mandelbrot had come up with the concept yet at IBM Research—but it would have applied. Because no matter how you magnify an angle, it looks exactly the same.

Now look at any angle in your current field of vision, or else imagine one—it's a very simple thing. Consider the exact point at which those two lines intersect. What, exactly, is the bearing area of that angle? Zero, right? It's a one-dimensional abstraction, a point. So over what surface, at the submicroscopic scale, does a knife apply its force?

The more I wrestled with this concept, the more confused I got. The bearing area of a perfectly sharp knife should be zero—nothing—*nada*. That gave the pressure ratio (force/bearing-area) a zero denominator, which made the ratio's value infinite. A perfectly sharp knife should cut through anything at all by its weight alone. It should slide smoothly to the center of the earth, where it would be effectively weightless, and stay there. Needless to say, a real knife does none of these things.

Jack Falstaff (later Susan's father) asked me what was troubling me one day, and I told him. "You're a moron," he said. Actually, as we were both friends and teenagers, he used a stronger term. "I don't care how sharp you make your knife; its edge area will *never* shrink to zero. Under sufficient magnification, it will look as round as a baseball bat." As long as the denominator of the pressure expression is positive, Jack explained, the whole expression can't be infinite. Problem solved.

Problem solved, sure—at least for 1962. But was it solved (as a true proof demands) for all places, cases, and times? Or was it solved only for the macroworld and the mesoworld?

It turns out the answer was the latter.

Twenty years ago, Drs. Gerd Binnig and Heinrich Rohrer considered a similar problem at their IBM European Research Laboratory in Zurich, Switzerland. They began to wonder if by some technical process they could not in fact make a blade so sharp, so perfect, that even at the nanoscale it would approach the ideal of perfection. Furthermore, they decided to make their perfect zero-area bearing not an edge, which is a straight line, but a point—a one-dimensional dot, a kind of immaterial position. What they wanted, in effect, was the world's most perfect needle. From such an origin, they reasoned, they could easily squeeze out individual electrons like individual bits of liquid from an eyedropper. Nanocurrents, a thousand times as small as microcurrents, could then be made to flow between this ultra-sharp point sensor and a material surface. These probe voltages (electron forces) and amperages (electron quantities) would be on an atomic scale. They would comprise not only actual electrons but "holes"—a quantum effect by which the absence of a negatively charged electron behaves somewhat like the presence of a positively charged antiparticle, called a positron.

A surface scanned by their new device, the two scientists reasoned, would influence the point probe in infinitesimal ways. By scaling up these tiny modifications, Binnig and Rohrer thought they could do what had never yet been done, what some distinguished scientists had gone on record as saying never *could* be done. They thought they could image matter at nanometer resolutions and at last see actual atoms.

The Swiss scientists were right. First they demonstrated their theory in the laboratory; then they created a workable instrument. This was the first scanning tunneling microscope, or STM. It and the variants that soon followed it—especially the atomic force microscope (AFM)—opened a window for scientists through which they directly saw, for the first time, the nanocosm. Later still, others developed the magnetic force microscope (MFM), whose sharp tip samples the magnetic fields that surround each atom. (How sharp is sharp? The tip of an MFM may have a radius of 20 nm or less.)

The STM principle was demonstrated in 1981 and won its inventors the Nobel Prize in Physics a scant five years later. That's lightspeed for a Stockholm review committee, which can take decades to decide that a discovery merits the Nobel. (The USA's Dr. Barbara McClintock did groundbreaking work on "jumping genes" in the 1940s, but her results were judged too startling to be true. Her Nobel came forty years later, nine years before her death. It's a good thing she hung on into old age. Dead scientists, no matter how great, are ineligible for a Nobel Prize.)

Science needed only months to recognize and apply the STM, and a few years to salute it: The instrument was that useful, revolutionary, and good. It did for nanoscience what the Zeiss optical microscope did for bacteriology—it launched an entire army of disciplines. Not bad for a Platonic knife blade.

POLLING THE NANOCOSM

Dr. Gianluigi Botton—who speaks Italian as his mother tongue, though his French and English are superb—is associate professor at the Materials Science and Engineering Department at McMaster, a small mid-continental research university, and holds a research chair in microscopy of nanoscale materials. As his titles imply, Gianluigi's explorations lie in the misty realm that's rapidly blending many different disciplines—in other words, the nanocosm. He's a chemist, he's a physicist, he's an atomic-force microscopist, he's a materials scientist, he's . . .

Well, he's high-class support staff, is what he is. As a resident electron microscopist in a university physics department, he's a helpmate for his colleagues. In fact, he's the guy who helps find what they think, hypothesize, calculate, pray, and hope like heck is in there. Gianluigi sets specs for the amazing, and occasionally cranky, multimillion-dollar instruments that peer up an atom's nostrils. He selects the devices; begs, borrows, or steals funding to pay for them; orders them; uncrates them; and sees that they're installed. Then he calibrates and maintains them. It's not surprising that he takes a paternal pride in everything that he and his fellow scientists afterward discover while using his beloved instruments. The kid who hit the home run? That's my boy!

Gianluigi is slim and easygoing, with a quick laugh and a fine sense of humor. Most of my interviewees are last-name quotes (*Smith* says); some are affectionate but respectful (*Dr. H* says). Gianluigi is that rarity, a first-namer. He loves his work and has a child's wonder at what "his" instruments are uncovering. I've seen the same attitude in a group of six-year-olds gathering rocks by the seashore: intense, unruffled, concentrated, serene. This time, though, I get the feeling that as Gianluigi collects his pebbles, the whole great ocean of Truth that lies before him is not going to stay unregarded.

Although the STM, AFM, and other new instruments are the glamour queens of nanoscience, there's still a lot of work to be done by an old, solid workhorse of micro-imaging: the transmission electron microscope, or TEM.

By the late 1930s, physicists had realized that electrons' schizoid ability to act like waves as well as particles made them useful for imaging small things. Thus the TEM goes back nearly sixty years. As a nine-year-old in 1955, I remember being shown a TEM by my uncle in his university laboratory. It was an established technology even then.

A TEM accelerates electrons with cathode-ray guns. It directs and focuses these fast electrons with magnetic lenses, then flings the electronic wavefront at a viewing screen. Phosphors in the screen thereupon fluoresce, re-radiating the electrons' invisible energy in wavelengths that people can see.

TEMs were the first instruments to image viruses, which had escaped the view of the best optical microscopes and still do so today. Science knew these tiny, biologically active structures had to be there. It took the TEM to prove them right. TEMs put a merciful end to the theoreticians' uncertainty—and one of the most grueling tasks in science is to make predictions that haven't been borne out.

Today, the TEM has morphed into the HRTEM, or high-resolution transmission electron microscope. In Gianluigi Botton's testing laboratory, one HRTEM fills a 200-square-foot room with benches, sensors, power sources, ancillary processors and instruments, and an eight-foot central tower housing. The tower contains the electron gun, image targets, and phosphor screen. At the HRTEM's controls I find an engaging young Japanese postdoctoral fellow with halting English and an infectious smile.

He gives me a great tip on my forthcoming research trip to Japan. The place to go to for basic nanoscience is Tsukuba, outside Tokyo, he tells me, and I make a note.

Gianluigi's HRTEM is all massive humming boxes and flashing lights and small, intriguing view windows. The monster I'm looking at needs only to be filmed in black-and-white to be at home in a B-movie from the Eisenhower era.

While Gianluigi's HRTEM uses principles first noticed and applied decades ago, and while it may look ungainly and old-fashioned, its appearance is deceiving. It's really an accurate, sophisticated nano-snooper with amazingly close tolerances in both its mechanics and its electronics. Its electric motors, which blend various types of instrumentation engineering, can shuffle a specimen up, down, and sideways to an accuracy of 3 nm. Then the motors lock the specimen in place while the HRTEM subjects it to a storm of investigative electrons.

Still, in the next room, things are very different. Here I find an atomic force microscope that is only eighteen inches high, takes up less space than a breadbox, and make almost no noise. But somehow its diminutive size and subwhisper silence make it more impressive—just as the stage actor who doesn't move attracts all eyes. The little high-tech instrument sits on a metal table that contains passive movement buffers and active shock absorbers to help isolate the AFM's tabletop from vibration. For the same reason, the whole table rests on a massive slab of solid concrete. Without all this protection, a mouse scampering across the floor nearby might blur the AFM's image. A wee-slip-of-a-thing grad student shuffling by in stocking feet could bollix the AFM completely. But here it sits, apparently motionless and, in the dim light, almost spooky. I don't see it doing anything until I detect a flat-plate video display that magnifies its field of view 600 diameters. The tiny ultra-sharp tip of the AFM sensor probe goes back and forth, back and forth, perpetually unsleeping. It stops only at the end of each micron-long traverse, then moves up a mere five angstroms (i.e., half a nanometer). It's scanning a surface, imaging individual atoms.

The devil in these details, Gianluigi admits with a sigh, is that oldest of demons—money. The AFM costs several hundred thousand dollars, and the HRTEM in the other room ten times as much. In 2003, eight years after that early HRTEM arrived, its replacement (already applied for) will cost three times as much again.

For the school that can scrape up the cash, however, the rewards that these machines promise are as big as their object resolutions are small. The new UHRTEM, Gianluigi tells me—U for ultra—will cruise the nanocosm. It will be able to do spectroscopic analyses "pixel by pixel," he says, extracting chemical information from the smallest possible part of whatever image it sees. "We'll be able to get data on electron bonds and energy states atom by atom," he tells me.

Gianluigi takes special pride in his sample-prep area. A specimen for the HRTEM looks like a tiny, shiny round cookie the size of a grain of rice. First it's mechanically milled, dimpled and thinned to about thirty microns—ten times thinner than the page you're reading, but (at a third of a million angstroms) ridiculously thick by nanoscale standards. Gianluigi then puts the sample into an instrument that bombards it with inert argon ions, slowly eroding it until its center is only 10–100 nm thick. Then into the HRTEM it goes.

"Electron microscopy is like political polling," Gianluigi sums up with a smile. "Your results are only as good as your sample."

CAPTAIN ATOM

I'm standing with Neil Branda, a young professor of chemistry at Simon Fraser University north of Seattle, on the floor of his virtual reality lab. It's a cool May day, and Branda has just got his motorbike out of storage after a winter of biblical rains. With helmet hair and a two-day jet-black stubble, he doesn't look like the holder of an MIT doctorate and, from several accounts, one of the brightest young lights in nanotech. At the moment he looks as if he's wandered in from the Downtown East Side to get warm. After a half-hour motorcycle ride through today's weather, he probably has.

None of this matters just now, because we're inside a molecule. Branda moves his hand. A tapering white virtual pointer swings through the air and stops with its tip on a nearby cluster of what appear to be colored tennis balls.

"There," Branda says. "That's a hydroxyl group. That increases the molecule's pharmacological activity. Okay, Brian, can you take us in? Stop just short of the hydrox."

Brian Corrie, the third man in this high-tech cave, nods his head. It's more than an acknowledgment. Delicate sensors on the man's headset detect his movement and relay it to a powerful Silicon Graphics workstation. The workstation simultaneously computes what all projection surfaces should display, and makes an adjustment. The image we see expands abruptly, flowing around and over us as if we'd dived into it. The tennis balls represent atoms—gray for carbon, blue for nitrogen, red for oxygen, white for hydrogen—so it's a patriotic drug that we're currently inhabiting.

"What's our scale here?" I ask.

Branda moves his pointer. "The H atom is about one angstrom in diameter, 10 percent of a nanometer. At this point, you and I are about two nanometers tall." He's flattering me. If Branda is two nanometers, I'm one point five. The guy is as tall as a basketball center, even when his height is diminished by a factor of one hundred and sixty-eight million.

The facility, called a virtual reality lab, really does seem like a cave. It's a small, dark cubic room ten feet to a side. Four of this unit's six sides—front wall, right and left side-walls, and even the floor—are image surfaces. These surfaces are filled by four big three-lens color video projectors, like the ones in sports bars. These units can refresh their projected image as often as thirty times a second.

The refresh rate varies with the complexity of the view. A simple object can be made to twist, turn, tumble, and zoom in and out as fast or as slow as you want, from roller-coaster velocity to a graceful pavane. An image that's really knotty—a hemoglobin molecule, for instance—takes up more of the workstation's RAM space and CPU time. Since each successive image takes longer to compute and display, we can't zip around it quite as fast on a nano-tour.

The molecule we're inspecting now is pretty basic, and Brian Corrie—a research associate with the New Media Innovation Centre in Vancouver—shows off his system's abilities. Suddenly I feel like I'm flying a 4-nm jet interceptor through this little molecule. We dip, dive, buzz nearby atoms so closely that we graze their outer orbitals, start and stop instantly, and turn on a dime. (I should say a proton—a dime is galaxy-sized by these scales, being over ten million nanometers across.)

Not only do we get acrobatics and color, but we get 3-D. The workstation projects two images alternately—left-eye field, then right-eye field, then back to left again—96 times per second. Each of us wears an $800 pair of goggles synchronized to this projection. Each eye's lens has an LCD shutter that can turn it from clear to opaque and back in about ten milliseconds. The goggles are slaved to the projection time-code, so they shut the left eye when the right-eye field is up, and vice versa. It's the effect you'd get if you could blink every 10 ms instead of every 40 ms, and do that for hours without tiring. Since the human brain updates its worldview every 40 ms, the interval of our own internal time-code, the workstation fools us into seeing two different images, left eye and right eye, at once. The brain fuses these two "simultaneous" images into a single 3-D picture.

The system works well, but it's complex. It's also cash-intensive, both to operate and to acquire. When I clumsily drop my LCD glasses on the hard floor, I can practically hear the beads of sweat popping out on Corrie's forehead. Or maybe it's my own. To cover, I ask him: "Anybody ever get sick when you jink around your images this fast?"

He grins. "I tell them what the IMAX people say. 'If you feel disoriented, close your eyes until the discomfort passes.' There's no real motion, so all your stimulus is visual. There's no conflicting data from your inner ear, like there is in seasickness or space sickness."

"What if someone feels sick and still keeps his eyes open?"

"The projection surfaces are scrubbable."

"Ever had to do that?"

He shrugs, causing the virtual images to spin violently and making my lunch threaten a return visit. "Once or twice."

Neil Branda has been pacing around the interior of his virtual molecule. "Okay, Brian, let's load the virus coat."

"The whole thing?"

"Let's start with one face. The triple-protein module."

Corrie doffs his goggles and ambles over to tap keys at a control console. The molecule clicks off, leaving the projection surfaces a spooky, Halloween-like shade: sort of an opalescent charcoal. After a minute, another 3-D image appears. This one is in three colors: yellow, blue, and green. Each color identifies a unique protein.

A protein is a linear sequence of different amino acids, each of which is in turn made up of atoms. For simplicity's sake, the workstation does not now show that level of detail. Instead the proteins appear as long, sausage-like shapes. The protein-sausages are kinked into wild convolutions—mad whorls and intricate zigzags. It looks like a Pipe Works screen saver on LSD.

"What," I ask politely, "are we looking at?"

"This is one face of an icosahedral protein coat," Branda says, staring at the image. "It's the armor plate of a virus that feeds on soybeans.

The full coating has sixty of these building blocks . . . Brian, give me the whole thing."

All of a sudden things get overwhelming. One tri-protein face was complex enough. Now we're treated to dozens of them together: tens of thousands of amino acids, millions of atoms. With the whole viral shell on display, the VR lab looks like it's giving us a close-up of an immense freeze-dried noodle packet. It's hopelessly confusing.

"Mph," Branda says, blinking behind his goggles. "That doesn't tell us much, does it?" Out comes the pointing wand. "Damn. Look at this, Brian."

Corrie looks closely, then zooms in. One of the sausages changes color along its length, which is impossible—no protein changes into another protein midway through a bend.

"I should inspect your data file," Corrie says. "That can't be right."

They confer for a few minutes. Then I ask: "Is this a deadly virus?"

"Only to certain plants," Branda says. "It's harmless to humans."

"What's its attraction?"

"Several things. First off, it's been thoroughly studied. Its shape and other properties are well characterized and its protein coat is stable. And as I say, it's benign."

"What's its appeal for you guys?"

"As a hat-rack for photochromic molecules. To immobilize them. No one's done this before."

"Where does the virtual reality lab come in?"

"In everything you saw." He grins lopsidedly. "Minus the file errors, of course. The VR lab gives us a sense of what this thing looks like. Inside looking out, outside looking in, and from all angles. Shape means almost everything here. It has a major influence on function. Once we have an intuitive feel for a protein's shape, we'll get a better sense of how to take advantage of that shape. How to use it as a molecular platform. Or even how to change the molecular function. And once we know that, we'll be in a better position to figure out how to tweak it."

"Tweak it?" The interviewer is the perfect straight man.

"Optimum drug-attachment points, possible shape modifications, places where separated charge could be stored. That sort of thing."

I find this very impressive. Most of the brain's processing power is in vision; harnessing that power for intelligent design and delivery of drugs, organic solar cells, and other new products seems a good idea. The VR lab is a complex tool, but it is also great for liberating the human imagination.

Still, Branda and Corrie hasten to tell me, their VR unit is no longer state-of-the-art. Even more advanced facilities are coming on stream. One of the best has just opened for business 500 miles eastward over the Rocky Mountains, at the University of Calgary. And that's not all. The great-grandma of all VR operations, one of the first and still the most advanced, is at the University of North Carolina (UNC) campus at Chapel Hill. This installation doesn't limit itself to virtual constructs: It shows the real nanocosm. More important, it doesn't just image things, real or imagined. It lets people get into the picture, moving things around at the atomic scale.

SCREECH AND RUMBLE

The North Carolina unit is more than a projection simulator. It's a multimedia miracle. First, what you see is really happening. Second, it's in real time. It happens not only the way you see it, but the instant that you see it. Third, the UNC nanomanipulator provides force-feedback that literally lets you *feel* the nanocosm, as well as seeing, analyzing, and messing around with it. Operators can prod a virus with a technologically mediated finger, thereby determining whether the bug's exterior is squishy (deforming plastically), springy (deforming elastically), rigid (stiffly resisting deformation), or all three things in various places.

Manipulation is the next logical step after imaging. Our primate ancestry makes it impossible for us not to muck around with what we detect: Mankind see, mankind do. Meddling lurks deep in our genome. Like many such things, we've become ambivalent about this. It may be a hardwired trait, but it's socially awkward, too. We love having this capacity to manipulate things, but act as if we don't. We dangle noisy, colorful toys above our babies' cradles to encourage the hands-on trait, then wonder why we have to tell our toddlers to look with their eyes, not their hands. They can't—it's preprogrammed. Next time your little darlings disassemble a DVD player, tell yourself as you count to a hundred: That's precisely what makes 'em human.

With the UNC nanomanipulator, we adult brats can now play in sandpiles whose grains are individual atoms. This is because the STM, the AFM, and other devices, collectively called scanning probe microscopes (SPMs), have been developed until they can move things around as readily as they image them. The forces that emanate from an SPM's super-sharp probe tip can be focused onto a single molecule, or even a single atom. Atoms bond to one another via electric charges. If an SPM probe zaps an atom with a voltage that exceeds its binding force, that atom forsakes its earlier attachments and sticks to the probe. Find another place to put your atom, take it there, hold the tip steady, and zap! Atomic resettlement. Seeing things at the nanoscale would seem miraculous enough, but now we can intervene to change things, too.

These handy effects were discovered by accident when SPMs were taken too close to the surfaces they were scanning, and scraped them. Since then, the science and technology of nanomanipulation have taken big strides. It's become routine to attach other molecules such as DNA to an SPM tip. This turns it into an active chemical probe whose action can be watched as it happens.

As an operator manipulates a scanning probe microscope, screens show her what she's doing, almost at the instant she does it. That's the principle behind the UNC nanomanipulator. It's a microscope—a nanoscope, actually—with very, very fine tongs.

It's a pleasant surprise for me to find that one of the key people behind the UNC nanomanipulator is R. Stanley Williams. After his UNC days, Stan Williams moved to California to direct basic and applied nanoscience for Hewlett-Packard Company. As such, he has become something of a Silicon Valley fixture. He constantly expresses the long-term vision that only a grand old man can present without being seen in the scientific community as a self-serving fringe artist. That's because he not only theorizes; he experiments as well. Stan pushes for increased NNI (National Nanotechnology Initiative) funding in the physical sciences and cautions against burdening the nanocosm with overhyped expectations and unworkable plans. But back in the mid-1990s, before he'd attained his Grand Old Man degree, Stan worked at University of North Carolina at Chapel Hill designing the nanomanipulator's human-machine graphic interfaces. These are vital to the operation of the nM—the abbreviation used officially by Chapel Hill for its nanomanipulator, and a nice play on nm or nanometer.

While the nM gives an astonishing sense of seeing and moving atoms directly, that's partly an illusion. This is still *remote* viewing, *tele*manipulation: seeing and handling things entirely by wire. Tiny fragments of raw data are amplified, noise-reduced, and presented in ways that make the most sense to a human user. (Of course, in fairness to Dr. Stan and his interface wizards, one could say the same thing about "direct" human senses.)

As we saw in Gianluigi Botton's lab, scanning probe microscopes such as those used in the UNC-nM move repeatedly back and forth over the scanned sample. Zip right—up a smidgen—zip left—up a smidgen—repeat. It's called rastering. Rastering is the same technique by which a glowing, pinhead-sized dot on a TV tube creates thirty sequential images per second, fooling the eye-brain system into seeing motion.

As an SPM raster-scans a surface, its perfectly sharp tip slowly and completely blankets the scanning plane—just as a theater line snakes back and forth to fill a waiting area. And as the SPM guides that super-sharp tip along its pre-plotted route, the tip constantly senses and records its distance from the surface below it. This information is relayed to the nM operator as it comes in. One of Stan Williams's interface techniques "tessellates" the image, or divides it into nesting black-and-white triangles. A supercomputer working with parallel architecture, forefather of the one that Dr. Simon Haykin wants to print and weave like cloth, massages this bland and boring gray-scale view. It shades edges and lightens certain corners so that vertical variations seem to leap off the screen. The added light areas, called "specular highlights," make the resultant views of a nanocosm surface seem rugged, detailed, and arresting, as if some imp were shining the beam of a powerful flashlight obliquely across the nanocosm's surface at a low angle. Users of the nM see craggy, detail-filled landscapes, as if they were flying a small plane low over the Rockies during a bright, clear sunrise in May.

Then the real fun starts. The nM—moving from sensing to manipulating—modifies the surface it's scanning. This is done in one of two ways. The nM can move things indirectly, without contact, using nano-voltages applied through the ultra-sharp tip. These electrical forces act on the electron orbitals of the surface atoms, pushing and pulling them about. Alternatively, the tip may physically press into the surface.

As a rule, modifications are made under computer control. This keeps a ham-handed human neophyte from damaging or destroying an SPM by (oops, sorry) inadvertently trying to shove a probe tip right through the sample it's scanning. In these machine-mediated cases, the SPM images the surface immediately before and after a surface change, letting an operator contrast the two appearances.

More experienced operators are allowed to control the SPM tip by hand. The scientists can't see the SPM tip touch the surface: They're moving things around with the camera lens, so to speak. But they can still get a feel for what's going on—literally. The sense used in these operations is not sight, but touch. The nM gives skilled practitioners a force-feedback option that relays and amplifies nanoscale mechanical properties such as hardness, resilience, and ductility. Operators sense everything about a surface's nanoscale characteristics except the scraping sound their fingernails would make on it. Like the VR emplacement, the nM lets nanoscientists develop an intuitive grasp of materials they're a billion times too big to work on directly. And like the VR lab, the nM literally gives people a feel for the nanocosm—not just as they visualize it, but as it really is.

I find it interesting that with the nM, the nanocosm exactly parallels the ultra-macrocosm. Today's nanoscience is redeveloping the same types of multiple sensing and variable-correlation techniques as astronomy has developed over the last thousand years. What science did for the stars, it now does for atoms.

The first astronomers were limited to noting and comparing celestial objects in only three properties: color, apparent brightness, and position. Then came reliable clocks, with their time-coding of observations; and telescopes, with their vastly magnified images. Next there arrived the spectrographs. These devices infer data about the surfaces of planets (and the guts of stars) from photons, at energies from gamma rays to weak radio frequency. For astronomers, spectrographs tell tales about objects that range from the supergalactic down to the merely colossal.

Today a visible-light image, on a traditional photographic plate or its digital equivalent the charge-coupled device, is only one of a wide array of astronomical tools and techniques. In an orbiting observatory, for example, sky position can be correlated with X-ray output over time. In this way earthbound astronomers can draw a clear map of a high-energy sky without once looking through a telescope themselves.

As with the macrocosm, so it is with the nanocosm. Nowadays a top-notch SPM can extract much more than images from the surfaces it examines. It can remote-sense temperature with high enough pixel resolution to paint an infrared portrait. It can read the forces jostling a given surface atom from any or all of its crowd of neighbors. It can map electrical properties such as resistance and conductance.

The nM at Chapel Hill currently displays these various readouts as multiple black-and-white images, placing them side by side on a big-screen graphic readout so an operator can simultaneously review all possible data in real time. And further improvements are coming. If nM planners have their way, color-coding and shading will soon let scientists see several properties mapped onto a single surface image. These overlapping data displays will instantly identify critical associations. Experimenters will, for example, be able to detect how well stress concentrations match electrical conductance variations along the edges and tips of physical features such as cracks and fissures. This will further one of the nM's main aims: to give nanoscientists an immediate sense of what correlates with what inside the nanocosm. Material scientists have for years theorized that the stress just ahead of a moving crack or dislocation varies as the size of the crack-tip's radius. The sharper the crack, the smaller its tip radius and the bigger its substance-splitting force will be. Or so goes the theory. Like all *elegant* theories, these calculations have been reasoned out from first principles and are logically sound. And like all *useful* theories, they have also been proven empirically effective so that engineers can confidently use them to design structures with acceptable safety levels. But until now, no one has directly seen stress develop as a crack moves through a substance at the nanoscale, or directly felt how the material responds at the nanoscale. In other words, no one has directly shown this useful, elegant theory to be *true*. Now this is happening. As Susan Falstaff told me with such excitement, science can now observe what it suspected strongly was there, but until today was forced to accept on faith.

The nM's designers hope to give its operators even greater abilities. By 2005, scientists may feel friction and adhesion (collectively called "stiction") and even hear the screech and rumble of the SPM probes as they slip, skip, and gouge across the face of the nanocosm.

THE NANOPRENEUR

Dr. Laura Mazzola did her undergraduate science degree in chemistry and mathematics at Kalamazoo College, Michigan. She moved to Silicon Valley, Northern California; developed a biomolecular sensor at "the Western MIT," Leyland Stanford University; worked for the NoCal biotech firms of Affymax and Affymetrix; and helped develop the technology that led to the first GeneChip, an Affymax product that tests DNA. Laura lives in Redwood City, California; I met her in San José, and we communicated by e-mail afterward.

I confess to bias here. As is apparent from our e-mail exchanges (which are reprinted here), I see in Dr. Mazzola everything a "nanopreneur" needs: education, experience, drive, humor, imagination, wit, and brains the size of Mission Control. She's the model of the person we need to propel the nanocosm from curiosity-based science to a strong force in the world economy. VCs of the world, keep tabs on this one.

I had no idea when Laura and I were writing each other that our letters would find their way into this book, with only a few added explanatory notes appearing in {ornamental brackets}. But after several tries at writing new text using these notes as reference, I've accepted that nothing done after the fact can approximate the lighthearted intensity of the originals.

```
From: Laura Mazzola
To: Bill Atkinson
Sent: Tuesday, May 21, 2002 3:52 PM
Subject: hello
```

Hello Bill,

Nice to meet you at Nanotech Planet earlier this week. I just thought I'd touch base to say "hello" and send you my summary and contact information. It was great to hear you remind the nanotechnologists to keep their predictions grounded in reality, not many people will make the effort to challenge the demigods in science. :)

Cheers,

Laura Mazzola, Ph.D.

```
From: Bill Atkinson
To: Laura Mazzola
Sent: Wednesday, May 22, 2002 10:44 AM
Subject: hello
```

Great to hear from you. Fascinating conference— one or two (OK, only one) people like you, with vast knowledge and smarts and a sense of humor (and great hair). Lots of nerds with noses to the nano-grindstone, and several of those delicious charlatans drifting about . . . The Secrets Of The Universe Unlocked, Only $450/hr! Even an earthquake . . . but hey, it's California.

As for bearding the demigods, I always think of Montaigne: "Be he upon never so lofty a seat, still a man sitteth only upon his own bottom." There's no one alive who can't be wrong. Even if he's right and I challenge him {and lose}, I've learned something.

I will definitely be in touch. Going to Japan in July (assuming agent sells US rights this week, hope hope) and probably to IBM Watson (NY) to see Tom Theis in August. But I'm planning a long chapter on biosensors; and if you're willing, I would very much like to interview you about your work. This is partly a business book, so I'm interested in that angle. Lots of lab rats to be found in nano, but not very many who know squat about commercializing the stuff. A lot of these guys couldn't run a roadside Kool-Aid stand.

B.

From: Laura Mazzola
To: Bill Atkinson
Sent: Wednesday, May 22, 2002 9:10 PM
Subject: Re: hello

Hey Bill, I look forward to our next encounter. I can blather away about biosensors any old time. It might even be useful!

Have fun in Japan—I spent 3 months in Tokyo, it was both exhausting and exhilarating.

Laura

From: Bill Atkinson
To: Laura Mazzola
Sent: Thursday, May 23, 2002 4:55 PM
Subject: Biosensors

Tell me more about your biosensors. What did you do? What are you doing? What have you got planned?

From: Laura Mazzola
To: Bill Atkinson
Sent: Friday, June 14, 2002 4:42 PM
Subject: Biosensors

Hi Bill,

I've now focused on becoming fluent in nano-biotech. Could lead to consulting in the near term, but my goal is to join or found a nanotech start-up within a year. I've offered to help the NanoSIG {Special Interest Group} organization get off the ground, trying to write up an executive summary of various bio-nano applications.

Biosensors: Molecular Velcro, that's what I built for my graduate research. It's called nanotechnology these days; back then it was called chemical force microscopy. Take an atomic force microscope tip, modify its surface with DNA (or proteins) and you now have a sensor for biomolecular affinity—at the molecular scale. I used it to probe other surfaces to detect and measure the force of DNA adhesion. I also worked at Affymetrix (& Affymax) in the early days to develop the technology for their high density protein and DNA arrays.

Laura

From: Bill Atkinson
To: Laura Mazzola
Sent: Saturday, June 15, 2002 1:39 AM
Subject: Re: Biosensors

Laura! Thanks for the reply—it's material like yours that lifts a pop-sci book out of the sesquipedalian and gives it zing. "Molecular Velcro" is a great term: it (forgive me) sticks in the mind.

Tell me this, though: How do you keep such DNA-mediated probes from gumming up within seconds? Can you do the molecular equivalent of degauss-ing the things, or washing the lint off the cel-lotape, so to speak, and revealing your half-DNA probe like-new again? Or do you have to keep reapplying fresh bait for each trolling pass? Why am I using so many horrid metaphors?

Good luck on all your ventures. I may be back in a few weeks asking for your take on a nano-business start-up. "Wisdom crieth out in the street, and Atkinson regardeth her."

B.

From: Laura Mazzola
To: Bill Atkinson
Sent: Sunday, June 16, 2002 2:30 PM
Subject: Biosensors

Hey Bill,

The total number of molecules defines the strength of adhesion. For my project, we are talking only 3-5 pairs of nucleic acid strands (and not fully "zipped") which resulted in adhe-sion in the range of a hundred nanonewtons—less

than the strength of the DNA-tip bond and much
less than the mechanical strength of an AFM. The
strength of adhesion can be modulated via the
kinetics of interaction, the total surface area,
and little chemical tricks like the salt strength
of the buffer. DNA, as you probably know,
requires a salty environment in order to
hybridize—remove the salt and the hybrid pair
instantly dissociates. {No, I didn't know that—
WIA.} I love nature, the molecular mechanics are
deviously clever and generally reversible.

So what do you think, does it sound like a valid
idea for a company? In fact, there are a few
already trundling down this path. It would be
perverse logic if I ended up going back to my
thesis work. They say you can never go back home
. . . :)

Laura

From: Bill Atkinson
To: Laura Mazzola
Sent: Sunday, June 16, 2002 9:04 PM
Subject: Biosensors

From my bitter, cynical male perspective it seems
a viable start-up is possible if, and only if,
your Ph.D. work can be converted into a solution
that others would pay handsomely for. And to say
yea or nay to this, I would have to know far more
than I do about the commercial surround for your
work. Elegant science: that's a given. The Subtle
And The Profound, sure. But also The Lucrative?
Dunno.

- B.

From: Laura Mazzola
To: Bill Atkinson
Date: Sun, 16 Jun 2002 21:40:51 -0700
Subject: Biosensors

Elegant science, yes, but I realized in grad school
that it was not the esoteric but the practical that
held my attention. Hence my summer project, to find
something both elegant and practical. Like a
Chanel suit! And yes, it will cost *beaucoup*—hope-
fully I'll find someone else to foot the bill.

See you.

LM

CHAPTER 7

WET NANOTECH

A CASE OF HUBRIS

TOTAL DIRECT world funding in nano-activity—in technology and science, both public and private—will likely exceed $5 billion by the end of 2003. Indirect funding will exceed this figure by an order of magnitude. The main reason for this rapid growth lies in a single subsector of nanotech. The life sciences, including genomics and biopharmaceuticals, are the biggest area of nanoscience R&D and the largest single source of nanotech funding. The biosciences are pushing ahead into nanotech faster than any other academic or commercial sector, even IT, and are central to the clear majority of nanotech start-ups.

Bioscience and biopharmaceuticals, it turns out, have one insuperable advantage in understanding the nanocosm: They have been working at the molecular level for a hundred and fifty years. In a sense *all* biotech is nanotech, and always has been. That gives it a big head start in nearly every subdivision of nanotechnology. Dr. Bryan Roberts, a California VC with a doctorate from Harvard, puts it this way: "Bioscience has worked in the nanometer size range for quite some time. It has a history of cross-disciplinary work, and a lot of its customers are cash-rich. Bioscience can readily combine these important advantages with its equally important knowledge and skill sets, and transfer them to nanotechnology."

This introduces a rich irony. A constant current in the early, speculative days of nanotech, and a trend still discernible today among nano-boosters, is the contemptuous disparagement of existing technologies. The boosters say that classical approaches to discovering and doing things, from bridge building to chemical engineering, rely on techniques that at the atomic or molecular level are so crude they're laughable. Even something as apparently marvelous as a semiconductor CPU chip, incorporating millions of microscopic transistors, resistors, and capacitors in a microcosm no larger than a postage stamp, is made using methods such as molding and photo-etching. Since these techniques shuttle around atoms by the quintillion, to the nanoboosters they seem (as they did to Richard Feynman 44 years ago) to have the subtlety of a ball-peen hammer or a double-bladed axe. Far cleaner and surer to treat atoms with the respect they deserve, say the boosters. Handle them one by one: Design and build exact structures *de novo* on the nanoscale. Nanotechnology, one booster wrote in 2000, would "snap atoms together like Lego blocks" to make whatever humanity desired. It would "replace inelegance with elegance in all forms of manufacturing."

Classical scientists who deigned to reply to the boosters objected that even a bit of matter too small to see with the naked eye might contain a quadrillion atoms—1,000,000,000,000,000. If a hypothetical (and still nonexistent) nanoassembler positioned atoms at the rate of one per second, it would take the age of the universe to accrete a speck of dust. Well, then, said the boosters, we'll use a geometric progression: a kind of engineering chain reaction. The first assembler will be made by classical, extra-nanotechnological means, molding and gouging countless atoms to do the work. This initial unit will then set to work assembling other units like itself, which will join in the same task; *und so unter.* On toward Zion!

As often happens, the artists had already foreseen this snowballing state of affairs. In a 1950s cartoon for *The New Yorker,* Ed Fischer shows two scientists observing a robot-assembly plant. As each new robot reaches the end of the line, it stands up and helps its fellow machines to assemble yet more robots. "My God!" wails one scientist. "Where will it end?" To the nanoboosters, the answer is clear: It will end in a paradise for technology. New technology. Nonclassical technique. The Golden Nano-Age. Farewell, axes and hammers.

Among all classical scientists, the boosters saved their most intense derision for chemists, whom they characterized as barbarians content to throw together random messes in the hope that something useful would result. Among chemists, biochemists—indeed all bioscientists—were the lowest of the low. Granted, their chemical reactors manipulated matter at the molecular level; but so did pig intestine. Great gouts of material were poured in, different kinds of gunk were extracted; but all of these sloshed together in huge, impure quantities. Humanity need no longer make things this way, by guess and by God. For the first time in history we could quickly find, or knew already, the exact makeup of any molecule involved in life—from complex proteins such as insulin and hemoglobin, to the deoxyribonucleic acids that constitute genes. Since all things alive or dead are nothing more than atom assemblies, all we had to do was cobble together whatever we wanted, beginning at the atomic level and working steadily up to the micro- and macroworlds. Piece of cake!

By 1990, at least to the nanoboosters, life itself stood naked and shivering in the glare of realism, devoid of all its ancient mystery. Now, the boosters said, gross approximation could yield to exactitude; for the first time, alchemy could be replaced by real science. The boosters expressed their derision in a verbal sneer. Even the most successful classical biosciences, they said, were just "wet nanotech." It was time to take nanotechnology out of the kiddie pool, towel it dry, and send it off to do adult things. Its first achievement would be to convert chemical synthesis into a subset of mechanical engineering. Atoms should be handled one at a time, with a Swiss watchmaker's meticulous care. Machine-phase matter!

When I looked into the nanoboosters' arrogance, I found it had some of its roots in their profession. Most boosters are engineers; and many engineers exhibit a streak of condescension to anyone outside their discipline. Sometimes this is quiet, a mere smug satisfaction at belonging to a self-perceived elite. But equally often this self-promotion, and the attendant disparagement of every nonengineering discipline, is stridently vocal.

While all this is glaringly obvious to outsiders, it is almost invisible to the engineers themselves. In a little book written in 1968, *The Existential Pleasures of Engineering*, New York civil engineer Samuel Florman examines his discipline's decline in popular esteem. Florman says the fall from grace began in 1950, ending a century and a half of high prestige for engineers. In the last half-century, public adulation has given way to public contempt. Florman looks at many possible causes, from the rising cost of civil engineering projects, to the disasters that occasionally befall great projects, to the uneasiness generated by advanced technology such as nuclear weapons. I find it striking that the one cause Florman does not investigate may be the actual one: the old Greek concept of hubris and nemesis. Icarus flies too near the sun and plummets to his death. Midas achieves infinite wealth and nearly starves to death when his touch turns everything, even food, to gold. A man wishes for eternal life, forgets to wish also for eternal youth, and shrivels with age until he becomes a cricket.

The overweening pride that the nanoboosters take in their concept of the new technology, which leads them to condemn all other ways of seeing and doing, seems to me a perfect modern case of hubris. Brian Leeners, a venture capitalist and software CEO in the Pacific Northwest, puts it this way:

> "Engineers are trained to believe in themselves totally, to think that no problem is insoluble by analytic logic. And not just by logic, but by *them*. In my experience most engineers have a rock-solid belief that given inclination and time, they could do anything better than you, me, or anyone. They're completely delusional."

The nanoboosters make two major errors. First, they fail to see that standard engineering principles cannot simply be taken from the macro-scale and applied immediately to the nanoscale. The nanocosm is totally, bizarrely different. Change in scale changes the thing. Second, wet nano-tech has emerged from being the boosters' whipping boy to become the dominant force not only in illuminating the nanocosm, but also in exploiting it. Fully half of all nanotech start-ups apply advances in the newborn bio-disciplines. Thus they rest firmly on a base of classical chemistry.

THE POWER OF WET

It turns out that wet nanotech (or nano-bio, or bio-nano—few use the formal term *nanobiosystemics*) had a lot going for it from day one. The biosciences constitute humanity's most sophisticated set of techniques for manipulating matter on the nanoscale. That they do so en masse, with up to twenty-eight orders of magnitude worth of molecules at a time, does not matter. Wet nanotech can still reach high levels of reaction pre-dictability and product accuracy.

Take something called "site-specific mutagenesis." This technique, which won the English-Canadian biochemist Michael Smith the 1993 Nobel Prize in Chemistry, lets biologists zoom in to a precisely predetermined location on a huge DNA helix, hold it immobile, and alter it by as little as a single atom. This is microsurgery on the molecular scale. If you inspect things from an IT viewpoint—as many bioscientists are doing, using another brand-new discipline called bioinformatics—you realize that Smith's discovery (invention, rather) is a molecular editing function. Search, find, delete, insert, modify. This is word processing for ACGT, the four-letter alphabet of life.

Similar methods developed by biochemistry and biotechnology have already put us on the road toward nanoscale control of the vital compounds called proteins. Proteins are a good example of wet nanotech's continual self-refinement over many years. On first discovery, proteins could be analyzed only by being "degraded"—broken into linear chains of simpler molecules called amino acids. But within a few decades, bioscience could describe proteins as three-dimensional assemblies. Wet nanotech can now see how proteins evolve in 4-space (i.e., how their shapes change over time). Today proteins can be unfolded, refolded, disassembled, reassembled, and tinkered with: in a word, engineered. But it's *wet* engineering, and it's totally unlike the nanoboosters' odd concept of dry, mechanical manipulation using little crawly machines. Back to the drawing board, Drex. Or else to the laboratory for the first time, to see if all those bright ideas really work.

The ascendancy of wet nanotech is no coincidence. Nor is it an example of a rich, powerful, mossbacked scientific establishment rolling over novel ideas out of sheer unwillingness to accept the new. Wet nanotech works, it's that simple. There are sound reasons why.

One reason we've already touched on: molecular vibration. At standard temperature and pressure (70° F / 20° C, 14 psi) an average molecule shimmies up to ten billion times per second. This isn't just a mild hum, either. It's a frenetically sweaty hip-hop. During each vibrational cycle, the molecule may displace itself side-to-side by half its width. And that's just for simple molecules. In more complex compounds, individual atoms can vibrate with additional, interior harmonics. The result is the completion of most commonplace chemical reactions in about one million-billionth of a second. Expecting a set of nanomanipulator calipers to grab something that fast-moving is more than hubristic: It's naive. Even the manipulator will do its own thermal dance. 'Tis here! 'Tis here! 'Tis gone!

Paradoxically, the same molecular vibration that makes a nanomanip-ulator unlikely is what gives classical chemistry, and its daughter bio-chemistry, much of their power. The atomic jitterbug underpins all life. Molecules in aqueous solution not only vibrate; they constantly adjust their mutual orientation. When molecules gather, they act as we do. Everybody eyes everyone else and wonders (if only briefly) what would happen if they got together. Thermal motion is life's great matchmaker, an endlessly active hostess who ensures everybody encounters everybody in every possible way. This is why perfect cold, though it preserves order indefinitely, is inimical to life. The only things that move at absolute zero are odd quantum-fluids like liquid helium; and these are dead, dead, dead.

Vibration not only makes a nanoscale atomic manipulator unlikely to function, it sounds the death knell of most nanomachines that have been imagined. The same forces that perpetually agitate living material, and the nonliving molecules that lie within it, would quickly batter an artifi-cial nanomanipulator into junk. The effect is called Brownian motion, after its discoverer. To a synthetic nanosubmarine cruising the blood-stream, the random molecular collisions of Brownian motion would seem like an avalanche of ten-ton boulders arriving head-on at a hundred miles per hour. A nanoscale artifact could be smashed in seconds.

Contrast this with the mass-production tactics used by classical bio-chemistry. Wet nanotech's approach is eminently workable. It succeeds because its nanovehicles are themselves solid. They are molecules, knit by strong electron forces until they are rugged enough to survive undam-aged in the troubled sea they inhabit. At the nanoscale most chemical environments, including the human bloodstream, are madly agitated and choked with pummeling debris. That's not an ocean that you enter lightly; but classical bioscience has learned to swim in it with ease.

COMMERCIALIZING WET
NANOTECH: APPLICATIONS

Wet nanotech has not contented itself with laughing off the nanoboosters. It has riposted by co-opting some of the boosters' own ideas, suitably adapted from fantasy to reality. Bioscience is currently spinning off a wide variety of technologies, ranging from medical therapeutics and diagnos-tics to entirely new molecular approaches to computing, that equal or exceed the functions that the boosters like to imagine for their nonexistent nanoassemblers. Throughout the nanocosm, wet nano is pushing into previously dominant areas such as information technology. And in certain applications, it is even getting itself dry.

Half-strands of DNA, for example, can easily be bonded to a water-free substrate. This anchors one end of each genetic ribbon, which waves in a sea of test sample like a stalk of kelp. Nanopreneurs like Laura Mazzola have combined this assay technique with the tips of scanning probe microscopes. The result is a wide range of exquisitely sensitive tests—trolling for individual atoms in Lake Nanocosm.

Inch-square biochips or "microarrays" have been invented to commercialize this probe technology. The biochips are subdivided into thousands of distinct subareas, each area containing millions of single-stranded DNA clones taken from one gene.

Here's how they work: Say you're a pharmaceutical scientist who wants to know the sum total of a new drug's effect on human kidney cells. You merely expose two identical biochips to genetic DNA taken from two different cell groups: one treated with a drug, and one (called the "control") left untreated. The biochips tell you within minutes which human genes are activated when the drug does its work.

A single biochip contains thousands of individual assays—one per "biel," or biological test element. (A *biel* corresponds to a computer display screen's pixel or picture element; the coinage is mine.) Each DNA strand on each biel is tagged with a tiny chemical marker—a molecule that fluoresces, or re-radiates light, when lit up by an external source. This source is often the intense, coherent light of a red or green laser.

A green-glowing marker can be attached to genetic material from treated cells, while untreated cells receive a red-glowing marker. When the test is complete, the biochip presents a complete, color-coded profile of which human genes the tested drug has induced to switch on.

The biochip process can be automated into an extremely high continuous sample rate, or throughput. One of the leaders in this area of nanotechnology is Virtek Vision International, located in Waterloo, Ontario (the Canada connection again). VVI, a corporate offshoot of a parent company founded in 1992 to exploit new ways of accurately driving laser beams, applies its expertise to a line of biochip scanners called ChipReaders. These read biochips an order of magnitude faster than competing hardware, using lower levels of illumination, and with the ability to spot five different types of marker molecule. Jim Crocker,

Virtek's CEO, cites limited competition in forecasting his company's fourfold growth to $100 million in yearly sales by 2006.

In mid-2002, the American Chemical Society (ACS) listed 110 companies directly involved in advanced commercialization of nanoscience—what ACS calls "nanobusiness." The list included manufacturers of materials, including nanoparticles; software and instrumentation makers like Virtek; capital sources; and nanoelectronics. The strongest single category was biomedicine: biochips, drug delivery, and therapeutics. In other words, wet nanotech.

At the first quarter of 2003, absolute dollar amounts of fully commercialized nanoscience were still small compared to long-established sectors such as IT. But this is to be expected. The disciplines involved are nascent. Much remains to be discovered about the weird realm where our everyday Middle Kingdom meets the quantum-atomic world. But thanks to the central role played by wet nanotech, *all* nanotechnology has acquired a strength beyond that of most other newborns.

PUSH AND PULL

It's interesting how science and technology find their way into the marketplace. Nonprofit R&D institutes, among them universities and government labs, are familiar with a concept called technology push. This is the effort exerted by those who originate knowledge to get their findings into the commercial world and earning money.

Technology push doesn't always work. For one thing, basic researchers often overestimate the importance of their raw results, or underestimate the sweat it takes to adapt them. A further task, one even more fearful than technical adaptation to the average scientist, is the need to find capital. In most areas of academic research, even investigators armed with an outstanding discovery must doff their lab coats, put on suits, and go cap in hand to the strongholds of investment money. It's a daunting challenge, especially to an egghead introvert who's more familiar with sensors and computers than with finance. After an initial half-hearted attempt, many academics resign technology push entirely and retreat to the laboratory, the only place they really feel at home.

But within wet nanotech, technology push has become subordinate to another effect called market pull. This is the force exerted by the private sector to locate promising knowledge outside its proprietary R&D labs, assess it, buy or license it, adapt it, and then package and sell it.

The past-masters of market pull are the big international biopharmaceutical companies. Their commercializing energy exceeds that of every other sector and discipline. I've talked to a brilliant team of mechanical engineers who in 1989 discovered how to make a diesel engine run so clean that its particulate and nitrogen-oxide emissions fall to one-fifth those of a standard passenger car. Their attempts to interest engine and truck makers in their product, while finally bearing fruit, read like a fourteen-year odyssey of heartbreak, frustration, and dead ends. These people had to push their technology through a stone wall.

What a contrast with the biopharms! The big drug firms have full-time staff whose only job is to scan the scientific literature for the first sign of a potential product. Their search is not limited to the formal journals, no matter how abstruse. It extends to reading discussion papers and work-in-progress run up on the Internet, and even to gossip overheard at scientific conventions. (In such cases, loose lips *launch* ships.) Scouts for biopharm companies tour working laboratories and are empowered by their pharma employers to have scientists sign ironclad contracts. In return for the scientists' present and future intellectual property (IP), the drug-company operatives offer tempting stock options, cash, and residual payments, all of which they promise to increase should the research lead to a marketable process or product.

The drug companies' ace in the hole is to offer what all scientists lust for: further research funding. Few researchers on earth can resist such temptation, even when the funder locks up the resulting knowledge in a steel cage. Sometimes this bait is dangled before the scientists' institutes as well as the scientists. This can be the most effective tactic of all, for as the novelist Robertson Davies observed, "Universities are unceasingly avaricious in a high-minded way."

For all these reasons, today's bio-industry is already in the lab, influencing what is studied, how, and when. Multinationals perch like ravens on laboratory windowsills, blank contracts in their claws. I'd be surprised if one bioscientist in a hundred, even a kid in grad school, hadn't considered the commercial potential of his or her work.

In today's bioscience, bridges open overnight between new facts and new financing. A report on some frontline laboratory advance in wet nanotech often reads like a company prospectus. By comparison, a book on cutting-edge research into soil compaction or seasonal bird movement is a gee-whiz exercise for most businesspeople—interesting briefly, or else not at all. Decry this if you will, but the intensity of market pull from the biopharms constantly advances wet nanotech—and hence all nanotech.

DENDRIMERS: TREES OF LIFE

Even basic research in wet nanotech has strong commercial implications. Take organic dendrimers, completely artificial molecules whose shape is "dendritic" or tree-like (Greek *dendros*). A dendrimer looks like an asterisk:

Or think of it as resembling, if you know the toy, a Koosh Ball—a floppy sphere of multicolored rubber threads, bound together at its center and easy for kids to catch.

Dendrimers were first created in 1981 by Dr. Donald Tomalia at the Michigan Molecular Institute. Back then they were curiosities, made for their own sake. But as often happens in science, they proved to have useful properties that no one, not even their creator, realized at first. Dendrimers, it turns out, can be perfect vehicles for the timely, accurate delivery of drugs. They can be fabricated in almost any size, up to that of large protein molecules such as insulin. But while proteins can fold oddly or change their shape unpredictably, dendrimers are cemented together with strong, rigid atomic bonds that do not hinge. More to the point as far as wet nanotech is concerned, dendrimers have a vast internal surface area (think of all the voids and surfaces on a Koosh Ball). These nanoscale nooks and crannies are ideal places to ensconce therapeutic drugs, which are often small molecules with strong effects on living systems.

Dendrimers may also act as vehicles for gene therapy (GT), transferring healthy genes to cells that lack them. In the past, a vector of choice for GT has been a gutted virus—a shell of viral protein armor whose genes have been scooped out, leaving room for the desirable gene. But gutted-virus vectors show an alarming tendency to be recidivist and return to their bad old ways. Once inside a cell, they may reassemble some or all of their genome and revert to type: You let in Grandma, but you find the wolf. Dendrimers are inert, and do not have this problem. You never have to worry that they'll revert to a nastier type.

Nanoscientists at Texas A&M are now investigating modified dendrimers that react to injections of a "trigger molecule" by dumping their therapeutic cargo of drugs or genes. The modified dendrimers (m-dendrimers) would, however, do so only after being absorbed into the target cells that need the medication. Thus an m-dendrimer wouldn't just make deliveries; it would be told by head office where, when, and how to drop off its load.

QDOTS AND NANOPARTICLES

The nanocosm is a single entity. It can be studied and exploited from a number of arbitrary viewpoints, but in essence it is a complete world, indivisible into different areas. Hence one goal of a nanoscience team is to be interdisciplinary—to set aside all traditional academic distinctions and function as something that transcends contributing academic faculties A, B, and C.

In a similar way, various commercialized areas of the nanocosm have a way of re-coalescing. It's as if the nanocosm is reasserting its unity even in the macroscale world of business. For example, after wet nanotech, the biggest area of the nanocosm being commercialized is advanced materials. But nanomaterials also figure strongly along the frontiers of wet nanotech. Much of the current economic strength in wet-nano start-ups stems from the ingenious ways in which they use material particles only a few nanometers in size. In the emerging nanoeconomy, wet nanotech and artificial nanomaterials have already begaun a beautiful friendship.

In tire manufacture, carbon black comprises countless particles of carbon in its amorphous allotrope, or noncrystalline form. Other allotropes include a slippery, layered crystal called graphite and an immensely strong crystal, transparent to both visible light and infrared, known as diamond.

However complex the structure of a car tire, it's a static item. Between birth and death, its only change is to wear out physically and break down chemically. Within wet nanotech, nanoparticles are assigned more subtle, complex, and varying roles. To achieve these, the nanoparticles are fitted to some sort of harness—an antibody molecule, say—that a living system can recognize, transport, and position accurately. Attached to such a harness, a nanoparticle goes along for the ride.

When used to coat synthetic bone implants in this way, inert polymeric nanoparticles can be persuaded to bond as tightly to growing bone as a natural bone cell will. This effect may lead to tissue grafts and synthetic implants that never loosen. Other nanoparticles, in this case tiny spheres of gold eight nanometers across, can be attached to cancer-specific antibodies. Injected into the body, they could seek out tumors and fix themselves to the tumor wall. After a few hours, the tumor's outer layers would have a wall-to-wall coating of solidly attached gold nanoballs. If the body were then irradiated with harmless levels of infrared light—what you feel on your palm when you hold it in front of a heat lamp—the nanospheres would absorb the radiant energy, heat up, and cook the tumor till they crippled, shrunk, or killed it. Since infrared penetrates living human tissue to a depth of several inches, this procedure could be performed repeatedly even on some deep tumors without the need for surgeons to break skin. Golden indeed.

The sexiest nanoparticles of all, and those finding the widest use in wet nanotech, are quantum dots, abbreviated *qudots* or *qdots* and pronounced "cue dots." These are tiny chunks of a substance, as large as twenty nanometers or as small as a single organic molecule, whose academic moniker is "semiconductor nanocrystals."

Qdots are fluorescent, re-radiating incident light in various colors. They're not just small, they're different. There's no analogue to a qdot on the macro- or mesoscale. This lets them serve a variety of medical uses. When bonded to antibodies that home in on tumors, qdots congregate in cancer cells. Then, in routine scans made during physical checkups, they reveal microtumors a few cells in size via fluorescence or lasing. Each color of qdot can be attached to a different type of biomolecule, instantly identifying it to a routine assay. If such techniques fulfill present promise and are successfully commercialized, they may literally spotlight a fledgling cancer years before its larger offspring show up on conventional scans such as X-rays. This set of procedures is very close to being a commercial medical technology.

Dr. Paul Alivisatos, a professor at the University of California at Berkeley, is a world leader in basic research about quantum dots. It's no coincidence that Alivisatos is also a cofounder of Quantum Dot Corporation (QDC), a private company spun off from his laboratory work. He typifies the rising breed of researcher-entrepreneur that's such a striking feature of nanotechnology, particularly wet nanotech.

QDC has already pushed one technology into the marketplace: a commercial test revealing tiny amounts of a specified DNA sequence. Like the biochip, this test starts with a "trolling bait" of half-DNA molecules—called cDNA because they complement another DNA sequence that's being sought. If the cDNA bait finds the complement it's seeking, in one ten-thousandth of a second the two strands line up and bond to create a full DNA strand. When this happens several million times along several million strands of bait, it triggers a massive, rapid clumping of qdots attached to the bait DNA. This abruptly changes the visible color of the sample. The test is sensitive, positive, inexpensive, simple, usable by unskilled personnel, and fast. It could be used to reveal pathogens such as HIV or hepatitis A, or even genes such as TNF (tumor necrosis factor) that the body produces when it fights otherwise invisible cancers.

THINKING MOLECULES

Wet nano is expanding so fast that in some areas it's going dry. A good example is molecular computing.

To my mind the most astonishing revelation of the 1953 Watson-Crick model of the DNA double helix is that at its core, a living gene is identical with information. Life, in other words, is synonymous with data. This isn't just an arresting philosophical notion, it's another smack in the eye for the nanoboosters. To date, their vision of nanocomputing has been SOS: Same Only Smaller. Shrink the circuit from microcosm to nanocosm, but keep unchanged the entire concept of the circuit. Tiny transistors, itty-bitty diodes, molecule-sized leads more fiddly than a cell phone's—all the architecture of a standard wiring diagram will remain at the nanoscale, only SOS. Eric Drexler even proposes a device like a nanoscale abacus.

Real science shows that nanotech need not limit itself to such paucity of imagination. There are ways of solving equations, or even of bypassing equations altogether, that totally dispense with electronic circuits. There may be no need to extrapolate silicon microtechnology downward, even in the short term.

"Why limit ourselves to electronics for computation?" asks Dr. Charles Lieber of Harvard University in Cambridge, Massachusetts. DNA, writes Lieber, is designed to store vast quantities of data, "and natural enzymes can manipulate this information in a highly parallel manner." As we've seen, parallel processing has emerged as a major strength of molecular computing. A parallel approach involves solving many small tasks at once, rather than plodding down endless one-way streets in the manner of linear, electron-based computing. When combined with analog architecture, this seems the way of the future for all IT.

A U.S. computer scientist, Dr. Leonard Adleman, has demonstrated that a DNA-based computer is a natural at solving problems that electronic devices find troublesome. And at the University of Wisconsin at Madison, Dr. Lloyd Smith has used biochips to speed up his computational work.

"DNA computing," concludes Lieber, "may ultimately merge with other types of nanoelectronics, taking advantage of the integration . . . made possible by nanowires and nanotubes." Alternatively, I suggest, DNA computing may dispense with SOS architecture altogether. We'll know within five years.

NANODIAGNOSTICS

At the University of Alberta, a laboratory team led by Dr. Chris Backhouse is using nanoscience to reduce the amount of medical test samples from milliliters (0.001 liter) to femtoliters (0.000000000000001 liter). The Backhouse lab even plans to go far below that threshold, down to single-molecule samples. The team's aim is cheaper, simpler, more accurate medical tests—and equivalent gains for the society that uses them.

Tests on large samples, explains Backhouse, tend to be slow and costly. "When we adapt macro test protocols to a nanosystem, we reduce costs and test time by orders of magnitude," he writes. "In medical diagnostics, these advantages are more than a matter of convenience. Reducing the costs of some tests by only ten times could permit nanosystem-based pre-screening for early cancer detection. The expense of existing methods gets in the way of such applications." Again, scale affects outcome: Change the number and you change the thing. Things are possible in the nanocosm that are not possible in larger worlds, at least not cost-effectively.

Backhouse and colleagues have already begun to develop various prototype nanosystems for medical diagnoses. Based on silicon microchips, these prototypes use genetic amplification and analysis. More to the point in performance terms, most of the tests show results in under two hours.

"Ultimately," Backhouse says, "we'll reduce test time to fifteen minutes or less. And we'll do it on smaller, fully automated instruments." At the moment, similar procedures are restricted to well-equipped macro-laboratories with a million dollars' worth of large-scale equipment. On average, each of these advanced tests takes several days of manual labor by skilled technicians, resulting in a unit cost of $400+. But, says Backhouse, "performance of the nanotests is comparable with or superior to the macroscopic counterparts." Necessary sample sizes have already fallen below the femtoliter level and steadily continue to decline.

Backhouse's lab has developed "an intense collaboration" with Micralyne, a firm that Backhouse calls "the world's leading microsystem foundry." The synergy in this alliance has already led to what Micralyne calls a "microfluidic toolkit"—an instrument for performing microchip-based bioscience protocols ten times less expensively than conventional equipment. By mid-2003, Backhouse plans to produce a handheld test device based on microships made by Micralyne. It will harness new miniaturization techniques to perform advanced medical diagnosis faster than any existing alternative.

Backhouse is also investigating nonmedical uses of this area of nanotech. These involve quantum dots called harmonic radar tags, which are nanodevices that transmit and receive radio-frequency emissions.

TWO JUMPS TO THE NANOCOSM

Dr. Chris Backhouse echoes many other nanoscientists, from Richard Feynman on. "There are," he says, "many paths to the nanoscale."

Research in biology, chemistry, physics, and engineering, Backhouse writes, will all soon converge at the nanocosm, as Feynman foresaw—in the Nobel laureate's famous phrase, "meeting at the bottom."

"Implicit in that statement," Backhouse says, "is that there are a lot of routes to the nanoscale, not merely one. Some of these new approaches will require basic research for decades before they lead to any practical applications." Does that mean that, for now at least, the nanocosm is merely an interesting set of speculations, more like a nanobooster's folk tale than hard science and technology? "Not at all. A shorter-term approach, the one our team is taking, is to reach the nanoscale by adapting existing microtechnologies. This creates applications that are specific and immediate. At the moment we're working to implement medical diagnostics. But our interdisciplinary approach to short-payback R&D may well discover basic effects that can then be embodied in other commercial technology."

Basic molecular biology, Backhouse writes, is ideally suited to on-chip implementation because by its very nature, bioscience is nanotechnology. Moreover, he says, "microchips can combine several different functions in a single geometrical area. They can perform cell sorting, separation of proteins, and polymerase chain reaction." PCR is a means of multiplying tiny quantities of DNA or RNA into detectable amounts.

The synergy between microsystems and biotechnology, according to Backhouse, can be applied almost immediately to many biomedical situations. "A lot of molecular assays are already being implemented as nanosystems, using the microfabricated, lab-on-a-chip technology. Other on-chip molecular assays are just as feasible." By molecular assays, Backhouse means locating one molecule among trillions. Finding the needle in a mile-wide haystack. Fast.

As Backhouse sees it, not only will new instruments perform these nanotech assays, the instruments themselves are about to undergo drastic shrinkage, perhaps as much as a million times by volume. The future, according to Backhouse, is microtech doing nanotech.

"The approach has several advantages," Backhouse writes. "First of all, it lets us approach the nanocosm in stages, rather than attempt a single immense leap." This, he explains, makes the microcosm a base camp between the middle kingdom and the nanocosm. Thanks to sophisticated and workable methods already devised to fabricate silicon microchips—molecular beam epitaxy, for example—we already know a great deal about how matter behaves at the scale of one micron (0.001 mm, or a thousand nanometers). This information doesn't comprise theory alone. It also represents much practical experience and know-how. Thanks to such work, sensors in the microcosm can readily pick up events in the nanocosm and relay them back up to us Brobding-nagians. For example, engineers at the Philips Research Center in Holland have devised virtual-reality graphic simulations that let chip designers "walk through" a proposed microcircuit as if they'd shrunk themselves to 1,200 nanometers in height. That's like Neil Branda's virtual reality lab seeing things a thousand times larger than a molecule—a micro VR lab, if you will.

Laboratories around the world are already making experimental one-off instruments based on chip technology. Some of these can detect quantities of matter under a picogram, or one billionth of a gram. In late 2002, Dr. Ash Parameswaran of The Micromachining and Microfabrication

Research Institute showed me a weigh-scale the size of a grain of rice. Its active site, an oscillator like a tiny diving board ten microns long, can detect the weight gain in a living cell after it gobbles fewer than a hundred sugar molecules.

Chris Backhouse anticipates much more of this two-stage effort to take mainstream science and technology down toward the nanocosm. He sees an especially strong role for microfluidics, the study of tiny amounts of fluid in correspondingly small reservoirs and tubes.

"Microfluidic systems," he writes in a review paper, "hold the promise of enabling inexpensive and rapid diagnostics by implementing analyses on a microscopic scale with dramatic improvements in speed and cost." That's the realm of the human body itself: the scale of the cell.

Microfluidic devices, Backhouse says, result "when you apply chip-fabrication technologies to produce microchannels in glass and plastic. Inside these narrow channels, you can manipulate very small amounts of fluid by applying electrical fields." Again, wet nanotech meets dry.

"You could argue that the revolution in consumer electronics was not due to the transistor," he says. "You could say it really came from integrated circuits that let us cram a million transistors onto a single chip. That's what really brought computers to every desk. We're doing an equivalent thing for nanodiagnostics—raising integration levels, making individual chips more powerful, and lowering costs. Microchip technologies can analyze single cells instead of vast cell populations. That's going to let us understand a lot more about life in general, and cancer in particular. Cancer is a challenge because not all the cells in a tumor are identical clones. Through microsystems and nano-bio, we hope to make big strides in controlling cancer."

ALL FOR ONE

The team that Backhouse leads is typical in nanoscience and nanotech research in that it combines different disciplines.

Broadly speaking, research is a constant tussle between what scientists themselves nickname "splitters" and "lumpers." As these names imply, splitters make increasingly fine distinctions, while lumpers connect things that once seemed separate. We have splitters to thank for the highly useful subdivision between chemistry and biochemistry. Lumpers surprised us a few years ago by showing that road traffic, like cold syrup in a tube, can be described using the same equations. Both traffic jams and "molasses in January" are subsets of viscose hydraulics.

The shock troops now storming the nanocosm come from the 101st Lumper Brigade. This is because our standard ways of doing science fall far short of what's needed for this strange new nanocosmic realm. A Ph.D., it's been said, learns more and more about less and less, until he knows everything about nothing. (By contrast, honesty compels me to admit, we science journalists learn less and less about more and more, until we know nothing about everything.) The splitters have done great things, but they cannot unlock the nanocosm. To understand even a fraction of what's going on down there, long-broken technical connections must be reforged.

To begin with, biochemistry must again talk with the inorganic chemistry from which it so recently and so successfully split away. These two subdisciplines, which have communicated only sporadically for decades, are not even the most extreme example of splitterdom run amok. A cell physiologist and a molecular geneticist, both of whom examine the living mammalian cell but have different disciplinary viewpoints, may occupy adjacent offices for years and limit their exchanges to the time of day.

Why is this? Mammalian cells are complex things, as intricate as big cities, so that in the short term specialization can pay benefits. Think of a new car model, which may have a whole team dedicated to a trunk latch and another team that designs only its exhaust manifold. As you become more farsighted, the rationale dwindles for such extreme specialization. A driver sees an auto as neither a bin full of parts nor a list of different systems, but a functioning whole. This type of conceptual integration is precisely what wet nanotech has realized it needs to do.

Dr. Steven Pelech, a university professor and CEO of the firm Kinexus Bioinformatics Corp., puts it this way in the U.S. bioscience journal *The Scientist*: "We are on the cusp of a dramatic change in research. It's like systems biology."

I find this a revealing image. IT has hardware engineers and software engineers. Both are splitter specialists working in distinct areas. Their work is synthesized into a functioning entity by a third type of knowledge worker: the systems engineer. You could almost define wet nanotech as systems engineering plus biology.

Not only must neighboring or nearly identical disciplines link up to explore the nanocosm, but vastly different technical species like engineers and oncologists, or software designers and infrared microspectroscopists, must now rub shoulders and share thoughts. In scientific terms, these types are as far apart as wombats and wolves. Nanoresearch, says an e-mail I received from a nanoteam in New South Wales, Australia, "may find that it is no longer possible to be neatly characterized by departments. The successful nano-institution will help, rather than hinder, the breakdown of traditional academic barriers."

Easy to say, hard to do. I cannot overstate how painful this is, nor can I pretend it won't get worse before it gets better. "Sometimes in nanotech meetings with a bunch of biologists," an engineer told me, "we waste hours in fruitless talk. Then someone realizes we've been using the same word to mean two different things, or two different words to mean the same thing. It can take half a day to agree on a single common definition." Friends, do you hate meetings? Then avoid the nanocosm. One university scientist in Iowa, who prefers that I not use her name, says her nanotech liaison committee reminds her of the chorus of demons who drag Don Giovanni down to hell. *No horror is too great for you! Come, there are worse in store!*

Hard though this work may be, it can pay off a hundredfold if its long-suffering protagonists stay at it. At the University of Alberta, Chris Backhouse says his main projects "link biomedical and clinical scientists, who use disease-based methods, with engineers who have the tools to design and manufacture devices. Then both groups talk with population-based researchers who look at the socioeconomic impact of the resulting inventions." The Backhouse program is typical of R&D in wet nanotech. It involves many different disciplines in both science and technology.

And the payoffs are immense. Scientifically, the basic discoveries of Backhouse et al. break new ground. Backhouse even speculates that the new diagnostics could lead society to classify health and disease "not by their supposed causes, but by the molecules associated with them." This could usher in a change as total as the Pasteur-Koch revolution of the 1880s. That scientific watershed moved medicine's focus from symptoms (fever, pain) to causes (bacteria, parasites, missing nutrients). Within a few years, a mere eye-blink in our great-grandparents' era, the new ideas of the mid-European microbiologists had generated intellectual property whose value it is hard to overestimate. This was the time of the first great pharmas such as Bayer. Microscience had led to macroeconomics. Wet nanotech promises a similar commercial breakthrough.

Then there's the matter of proteins. This is an area in which the United States, driven by large private companies and by the U.S. National Institutes of Health, is emerging as a world leader. Both scientifically and commercially, the stakes are high. Proteins are the master molecules of life. Besides being vital to all life processes, proteins make our skin elastic, knit our wounds, carry oxygen to cells inside our bodies, give hair its beauty, and lend bone its strength. There are thousands of known proteins, and probably fifty thousand or more still to be found. Nanodiagnostics will give us further insights into how proteins work—how their strange shapes and critical functions develop from information locked inside the human genome. Wet nanotech will also test the revolutionary hypothesis that a living system's environments provide some of the key information that it needs.

To date, a big stumbling block in the new discipline of proteomics (*prote*ins + gen*omics*) has been the minuscule amounts and concentrations at which proteins naturally exist. Like other valuable currencies, proteins are in short supply. Often the body needs to synthesize only a handful of molecules of some important protein such as thyroid growth hormone or interleukin-B. Its immediate task done, the body then breaks down the protein and stores it only as genetic information until the next time it's needed. It's rather like keeping a cake on file as a recipe, in between the times we want to eat a cake. But proteomics can't very well study what isn't there.

The big recent advances in genetics stem from a technique called polymerase chain reaction (PCR). This lets scientists multiply tiny bits of genetic material until they have relatively large quantities of it to study by macroscale means. But proteins have so far resisted such convenient multiplication. If we want to suss out proteins, we must do it on their terms, in whatever tiny quantities nature provides.

Enter wet nanotech. Using techniques collectively called soft lithography, proteins produced by living systems can act as templates that establish extremely precise and regular nanoscale arrays. Using these natural arrays as molds, known polymer technology can then make titer dishes that contain not one reaction site, but literally billions of them—each only 10 nm or less in diameter. In this way a single test surface can support upwards of a billion separate biochemical reactions per square inch. Swiss companies excel in these "nanomold" applications, which may soon provide a ten-second home pregnancy test. Dr. Angela Belcher of MIT and the University of Texas is experimenting with using the protein coats of bacteriophage viruses as templates for artificial objects such as nanotiter dishes.

Routinely analyzing tiny quantities of proteins and other medically important substances will be a giant leap for nanodiagnostics. If wet nanotech rises to the challenge through labs such as Chris Backhouse's, new medical applications from the nanocosm will profoundly transform us, our health, and our society. Unconvinced? Try your own thought experiment: Consider what micron-sized nano-instruments might do for the world's strained health-care systems. Bedside diagnosis based on nanotech could make lengthy, labor-intensive lab tests things of the past. Being less expensive, nanotests could lower the cash burdens on insurers, their individual and corporate clients, and the governments who pay for public health systems. By being more reliable, nanotests could potentially cut malpractice awards, reducing insurance premiums and thus the funders' costs. Better diagnostics could also raise success rates for medical treatments, while at the same time increasing doctors' confidence that they've chosen the best course.

Now go further. Imagine what would happen to public health around the world if even *some* cancers could easily and inexpensively be found in the time it takes to drink a cup of coffee. Here's my assessment. Locating a tumor when it's six weeks old and eight cells in size, instead of six years old with eight hundred million cells, could pare society's treatment costs by fifty or a hundred times. Healthier populations, unfrazzled doctors, hospitals underused instead of overcrowded, lower taxes, manageable public and private costs—even when limited to diagnostics, the promise of wet nanotech is immense.

Systemically, all these transformations will begin with multidisciplinary teams. The Backhouse group, for instance, has already developed links beyond local industry and academia to nodes all over North America and Europe.

"The research strengths arising from joint perspectives," Backhouse writes in an e-mail, "provide a unique interdisciplinary environment unmatched elsewhere. This environment offers novel insights, new technology, and substantial economic benefits. I see our approach as a model to be followed by others who want to develop nanotechnology and adapt it to a microsystem."

THE NEXT FIVE YEARS

In the intermediate term, say to 2008, a synergistic approach to the nanocosm promises achievements far beyond the scope even of medical diagnostics. Two Australian scientists, Dr. Vijoleta Braach-Maksvytis and Dr. Burkhard Raguse, coauthored a 2001 report on nanobiosystemics (i.e., wet nanotech) for the Asia-Pacific Economic Cooperation Forum (APEC). This influential organization includes not only "Oz" and New Zealand but also the United States, Japan, Korea, Canada, Indonesia, Singapore, Taiwan, and twelve other nations. The Australians predict that as well as supporting revolutionary diagnostics, wet nanotech will create nanoscale machinery, novel manufactured goods, and even artificial human organs.

There is a sound base for these assertions. According to the Australians, "[One] example of looking to nature for basic concepts is self-assembly. Invariably, biological systems put themselves together through the interaction of simple subunits. These organize themselves into ever more complex structures, without lithography or other external structuring."

The APEC authors recommend that human science mimic this type of natural self-assembly. In so doing we could create products such as protective surface coatings that, like healthy skin, heal themselves whenever they are torn. Nanoscientists in Germany are exploring this.

Scientists in France have used the self-replicating abilities of DNA to create solid silver wires less than three nanometers in diameter. This, they write, "opens up the possibility of using different types of DNA to wire up nanoscale circuits in two or three dimensions."

When these new circuits are commercially produced—grown, rather—many of them will be optoelectronic, using both electricity and light. Computation will be electronic; switching will be optical. NASA's Ames Research Center has already demonstrated an optical nanoswitch. The "flop," or reversible device to shunt current, was a donut-shaped molecule isolated from ancient organisms called archaebacteria. (These primitive bacteria are also called extremophiles because they thrive at temperatures above the normal boiling point, together with the tons-per-square-inch water pressures found at the bottom of the abyssal sea.) NASA's discovery paves the way to molecule-sized read-write devices—perhaps the first true IT to come from the nanocosm.

Finally, wet nanotech could power a long-sought goal for IT: molecular memory. Scientists at the University of Cambridge, England, have molded a naturally occurring protein called bacteriorhodopsin (bR) into a simple IT memory device. Unlike today's computer memories, which encode data on flat surfaces, this bR-mediated memory is as three-dimensional as the human brain. The Cambridge discovery opens the door to memory densities that are orders of magnitude beyond anything now available.

ONE BETTER ON LIFE

"I'm a chemist," Dr. Dipankar Sen tells me. "Not a biologist. Biologists seem content to observe how nature works. But as soon as I discover something, I want to meddle with it. I want to see if I can improve on it, at least in human terms. So I tweak it here, there, and everywhere."

We're sitting in a coffee shop at the corner of Davie and Granville Streets in downtown Vancouver. It's a warm, sunny day, and girls in structurally impossible tank tops saunter by. I wonder how I'll pay attention to my interview; then Dipankar starts talking, and I'm caught up in his ideas.

We're meeting here because Dipankar's nearby condo is, he tells me, "an unholy mess." He's been traveling so much that he hasn't had a chance to clean it. I've managed to snare him for a couple of hours in a three-day layover between a long visit to a mathematics institute in France and a nanotech conference in Sapporo, Japan. Often, Dipankar says, it takes him a couple of minutes after waking up to remember where he's been sleeping. You're in Vancouver, I tell him, and he makes a face. "Not for long," he says.

"I work on the edges of things," he tells me, sipping his latte. "The odd, gray areas that lie at the boundaries of current disciplines. If you do this in science it may lead to your ostracism, or at least to your neglect. You can pursue such a course only by doing work that is absolutely rigorous. Otherwise you leave yourself open to criticism."

Dipankar Sen originally trained as a biochemist. But in the last ten years he has become so fascinated by self-assembly and other natural processes that he now considers himself a nanoscientist. As a biochemist *manqué*, Dipankar considers DNA (the genetic material) and RNA (which transmits DNA data to cellular microfactories called ribosomes) "almost the same thing . . . . But nature has chosen DNA to be the stuff of genes because DNA is more robust, more chemically stable." Dipankar calls his approach "escaping the ACGT prison"—referring to the four nucleotides adenosine, cytosine, guanine, and thymine that constitute natural DNA.

"In aqueous solution," he tells me, "and at the nanoscale, DNA exhibits a lot of useful structural properties. A few scientists who want to get nonbiological materials to assemble themselves are looking to DNA to show them how to do this. I'm part of that group. I'm in Vancouver, some of us are in Europe, but most of us are based in the USA."

There's no physical reason, Dipankar says, why Watson and Crick's double helix can't shake off its stuffy image and learn to act more imaginatively. But this will take direct and purposeful human intervention. In nature, he tells me, any wonky variant of a standard double-helix structure is blitzed by compounds such as helicases. These are families of editing enzymes that cruise newly replicated natural DNA and edit out all products stemming from unorthodox replication. Dipankar, however, doesn't consider himself constrained by nature's draconian rules.

"My interest is partly intrinsic," he tells me. "Like a bioscientist, I'm curious about DNA. But as a nanoscientist, I'm also fascinated by what happens to DNA when we extend it beyond its natural parameters. In the natural world within the cell, DNA doesn't catalyze any kind of unorthodox biochemistry. But now we're making it do so. The final component of my interest is biomedical. I want to see if new DNA molecules with novel biochemistry can lead to new types of treatment for disease." Fixing what nature hasn't got around to fixing yet? "Yes, if we can."

A third area of interest for Dipankar Sen lies in nontraditional computing. "In speed," Dipankar tells me, "DNA has nothing on silicon. But DNA computation lets you be massively parallel. Sure, each DNA computation is simplistic. But it can still let you perform a quadrillion separate calculations per second."

Sufficient quantitative change leads to qualitative change: Crank out enough operations per second, and pedestrian computation can suddenly stand revealed as something rich and strange. By such means—that is, speed alone—the linear-digital approach of modern computers has parlayed conceptual idiocy into some passable imitations of human brainpower. With enough flops per second, one-two-*duhhh!*-three can create a world chess champion. With enough individual cells, however dumb, a DNA-based computer could write a novel. Come to think of it, that's how Tolstoy's *War and Peace* was written. No single neuron in Count Leo's brain had any smarts to speak of. Working together, they turned out a classic.

"A silicon chip is essentially two-dimensional," Dipankar adds. "Artificially modified DNA gives us the possibility of computing hardware that's truly 3-D. At a blow, we can compute at the nanoscale. We can do so with hardware as massively parallel as the human brain, and with component densities a billion times greater than we have now."

THE FIRST KILLER APPS

Many venture capitalists believe that wet nanotech will provide the first nanotechnology applications to achieve sustained commercial success. Candidates for these "first killer apps" include qdots, as well as new biosensors hardly larger than the molecules they detect. Some of these probes can sense as few as ten target or "analete" molecules. U.S. companies such as Nanosys and Molecular Nanosystems lead in this field. In their bioelectric sensor designs, an analete molecule completes or interrupts a circuit that's based on a single nanowire. This not only betrays the molecule's presence, it provides quantifiable results, showing both *what* is there and *to what extent* it's there.

Wet nanotech is also completing the movement of bread, wine, and cheese manufacture from the empirical to the scientific. These biological substances are now almost as engineered as cell phones. Process controllers can detect, interrupt, and adjust the exact characteristics and outputs of any biological or mechanical subsystem in a bakery, winery, or cheese plant.

Killer apps from wet nanotech can also apply to many nonbiological systems, says Dr. Bryan Roberts. He's a general partner with the California venture-capital firm Venrock Associates, and his specialty is seeking out and financing applications from wet nanotech. Under his direction, Venrock has bankrolled U.S. companies such as Caliper Technologies (expertise in nanofluidics), Illumina (nanoscale optical fibers), and Surface Logix (micro- and nanofabrication). Commercial activity in wet nanotech, Roberts tells me, is starting to surge.

"There's a fundamental agreement among large corporations, governments, and scientists," he says, "that many different areas and disciplines are coming together in nano-bio [wet nanotech] . . . . Previously separate fields are experiencing a strong convergence." In fact *all* of nanotechnology "is, now at least, more like biotech than like IT." At the same time, Roberts warns against overoptimism: "We don't want this [wet nanotech] to turn into another Next Big Thing bubble."

Wet nanotech start-ups, Roberts believes, have two main challenges. One is that they need lots of money from the get-go: "They're highly capital-intensive." The second problem is the shortage of qualified people—"especially people who combine the relevant scientific and technical knowledge with some entrepreneurial experience." This constraint is not unique to wet nanotech. Employers voice the same complaint in every knowledge-intensive sector in the world. But the shortage of skilled personnel in wet nanotech may be the worst in existence.

Someone who does fill the bill for an alpha employee in wet nanotech is Dr. Angela Belcher. Like many of the top people in U.S. nanotechnology, she's mobile, having moved her twenty-person R&D team in 2002 from the University of Texas to MIT. Austin-to-Boston is nearly 2,000 miles as the crow flies, and having come from a family that for nearly 200 years has been Texan, Dr. Belcher concedes the move was a bit of a wrench. But as the French visitor Alexis de Tocqueville observed with astonishment in the mid-1800s, Americans have a habit of pulling up stakes and relocating at a moment's notice. The lure of the unknown, the mystery of what's in the next valley, is just too great. "Americans," said the New England poet Stephen Vincent Benét, "are always moving on."

While it's made her peripatetic, Dr. Belcher's personal frontier isn't geographical. It's basic nanoscience that's quickly leading to new technology—in other words, wet nanotech. Angela Belcher's work has attracted attention and financial support from large and wealthy firms such as DuPont and IBM. Belcher's expertise in self-assembly at the nanoscale, writes Small Times analyst Candace Stuart, "could be potentially groundbreaking for both industries. IBM is in a race to miniaturize electronics to meet demands for faster, smaller, and more powerful computers. DuPont is pressed to find ever-better materials for customers ranging from paint manufacturers to nylon factories."

Listening to Dr. Belcher summarize her work is astounding. The woman buzzes with energy. She talks rapid-fire, as if compressing as much data as humanly possible into the minimum length of time. The ideas spill out one after the other, and they are packed together like sardines in a can.

"We've managed to marshal bacteriophage viruses pretty well," she says. "We've made cellophane-like films that comprise 99 percent virus, interrupted at intervals by qdots. Here's an example . . . . As you can see, the viral spacings are regular." She holds up a clear, colorless thin-film and shines her laser pointer through it. My jaw drops: There's "regular" and then there's *this*. The elements in Dr. Belcher's thin-film are so perfectly spaced that the pointer's ruby light generates diffraction patterns on the wall beside us. This is precision to a degree that would put most microengineers to shame.

"The virus is harmless to humans," Dr. Belcher says. "But immobilized in the film in this way, it stays biologically active in plant bacteria and loses only 10 percent of its infective power each year at benchtop conditions." A single square centimeter of film holds a trillion individual viruses, Dr. Belcher tells me. This seems the equivalent of stating that the film has an extremely high storage density and persistence of information, encoded in the nanocosm. It could be the key to molecular memory by another name: viral ROM.

Dr. Belcher is an expert in the links between bacteriophages and IT. Some of the viruses she has found and replicated have low-molecular-weight proteins, called peptides, that selectively bond with gallium arsenide, a substance almost universally found in microchips. And not only with GaAs, but with any one of its various crystalline forms that you'd prefer to isolate. Dr. Belcher has developed a whole toolkit of viruses that will locate and link up with whatever type of GaAs crystal you require.

This is commercial dynamite, because GaAs is probably the single most interesting material to the makers of microchips. Used as a dopant (trace alloy) in silicon dioxide, it converts simple silica to the all-powerful semiconductor material that's at the heart of nearly every computer in the world. Compounds that discriminate between various GaAs crystal configurations, and between gallium arsenide and silicon dioxide, are of intense interest to computer manufacturers. This is especially true when those compounds are not just binding agents but also growing agents, capable of getting key atoms and molecules to self-assemble with the precision of military drill teams. Dr. Belcher has even isolated "graphic-specific peptides" that bind only to carbon nanotubes. Hence her group's high level of support from IBM.

IT'S EVIDENT: Wet nanotech will soon be coming to a computer, hospital, and factory near you. The revolution is already underway. Advances in wet nanoscience and technology, together with more flexible ways to unite scattered fields into a single powerful nanodiscipline, will by 2007 have answered strong market pull with $500 billion worth of products yearly. It's as certain as the sunrise.

CHAPTER 8

FULLERENES, BUCKYBALLS, AND HUNDRED-MILE ELEVATORS

SOOT

THE BIGGEST news in nanotech today is grunge. Filth. Crud. The soft, black, fine-grained stuff that smears your hands when you clean a car engine, a kerosene lantern, or a wood-burning fireplace. It's found almost anywhere an organic substance has been burned, and has an Anglo-Saxon name unchanged in centuries: *soot*. Strangely enough, new products based on this humble substance have spun off one of the most commercially advanced subsectors in today's nanotechnology.

Soot has been part of human life since we harnessed fire. Some of us streaked our faces with it when we went to battle. Some of us (e.g., football players) do so today. Are those black streaks beneath a QB's eyes really there to cut glare? Aren't they there to make him fearsome to the enemy tribe?

Soot even goes back beyond fire. It antedates humanity. It's 100 percent carbon, atomic number 6, and life on earth (which is all the life we know) is based on it. Soot is literally in our genes.

Soot also underlies a lot of current nanotechnology. Carbon atoms are gregarious little guys: They stick to almost anything. Carbon even attaches to itself. It *particularly* likes to attach to itself, using something called a covalent bond. Just as couples bond by sharing an interest, atoms link up by sharing an electron. Covalent bonds involve not one, but two electrons. That double dose of cement makes them fast to make and hard to break. Bowling *and* stamp collecting, my dear: how lucky we are.

As well as being gregarious, carbon is relatively commonplace. The element goes far beyond that omnipresent soot. Look out your window: Every growing thing you see—grass, shrubs, forests—is half carbon by weight. The ratio becomes even higher when living plants die, letting noncarbon elements such as oxygen and hydrogen escape into the air.

Soot is very, very stable. It can stay the way it is forever. And most of it is nonradioactive. The carbon isotope C_{14}, which *is* radioactive, exists in small quantities and lies behind a technique known as radiocarbon dating. But C_{14}'s half-life—the time it takes for half a given chunk of it to emit an electron and become boring old nitrogen-14—is only 5,570 years. That's an eye-blink compared to the fourteen billion years that a chunk of thorium-232 takes to transmute half of itself into a stable, nonradioactive substance. Carbon knows when to leave the party, which may account for its high success with other elements. (Of course, if nobody else is available there's always the covalent, solo job.)

The chemistry of carbon is central to life. Imagine six carbon atoms in a long string, every other atom snugly coupled to its neighbor with a covalent bond. Now imagine bending the string around so that the two atoms at the ends of the string can bond. This turns the string into a six-sided loop called a benzene ring, which is the lynchpin of organic chemistry.

A high percentage of drugs, both old and new, are based on the benzene ring. This is because each angle of the hexagon provides an attachment point for medically useful stuff. Each covalent carbon-carbon bond, which chemists abbreviate C=C, works like a climber's clip ring. It snaps onto almost any molecule, even a molecular fragment. Any such molecule may be pharmaceutically active.

And yet the drug industry's reliance on the benzene ring for drug design has had difficulties. While each atom in a benzene ring is firmly bonded to both its neighbors, these bonds are not rigid. They act as hinges and are floppy enough to let the benzene ring deform. Uncounted drug designers have gone prematurely bald, tearing out their hair at the benzene ring's plasticity—its cheerful willingness to let itself be yanked out of shape. This is a problem because a pharmaceutical molecule is like a protein: how well it works depends on its shape. If a benzene ring at the core of a medication flops from circular to elliptical it may reduce, block, or distort that medication's function. In a sense the drug, like other vehicles, goes nowhere with a flat tire. If we're going to use soot to ease discomfort, alleviate symptoms, and compensate for deficiencies, could we please have a *stronger kind* of soot?

Enter carbon nanotechnology, with a set of molecular allotropes called fullerenes.

LESSONS FROM MONUMENTS

Unknown to drug designers, that last vexing question was answered by a brilliant, eccentric U.S. engineer named R. Buckminster Fuller sixty years before it was asked. Bucky, as he was called, thought so far outside the box that he redesigned the boxes. He started where a genius should start—with the basics. It wasn't enough for him to ask "How?" He took inquiry to the next level: "Why?"

The questions that Bucky asked himself, purely out of selfish, impractical curiosity, he grouped under the humdrum heading "close-packing of spheres." The topic intrigued him when he saw the vast variety of ways that America's regional architects had stacked piles of ornamental cannon balls around Civil War monuments and the entrances to civic buildings. There were cylinders and pyramids, cubes and cones. And seeing them, Bucky thought: How many ways could I fit together a given number of identical solid balls? Which ways would be most efficient? Which would use the minimum volume? What shapes and patterns might arise from my rules and techniques of assembly?

As he turned his brain to this problem, which was far more complex than it seemed, Bucky Fuller made a groundbreaking discovery. Once he'd isolated a close-packing pattern for solid spheres, he imagined all those spheres turning transparent. A further leap of Bucky's fertile imagination connected the center of each vanished sphere to its closest neighbors with thin, light, rigid struts. When he left the imaginary realm of thought experiment and duplicated these strut patterns in real materials, Bucky found that his newly discovered strut patterns, which were in effect the material tracery of his imaginary spheres, were both unconventional and efficient. In fact, they were extraordinarily efficient. These patterns were unlike standard engineering designs. They used no traditional elements such as columns, beams, and trusses. They were like nothing on, under, or above the earth—or so Bucky thought. But they worked like gangbusters.

One of the new designs was a type of flat, hollowed-out panel. It was as rigid as a solid plate but many times lighter. Everything was removed except thin bars of material. Each of these bars contained one of the internal force lines that, analysis showed, lay invisibly inside a solid plate. Bucky called this design a "space frame." Another pattern was a curved shell, which could be either a complete sphere or else part of one. Bucky called his struts "geodesics" and his shells "geodesic spheres." If the spheres were incomplete, he called them "geodesic domes."

Some designers think there is no theoretical limit to the size of a geodesic dome. Even using everyday materials with conventional strength and structural efficiency, you could roof in Tokyo, Paris, or Manhattan. The U.S. Pavilion at Expo 67 in Montreal was a single geodesic sphere sixty yards in diameter, which housed its exhibits without a single interior support. I was twenty-one when I visited this dome. From the inside it soared overhead magically, as apparently weightless as the sky. Artistically, architecturally, and scientifically, it was a triumph.

Now, barely a generation later, there's more—much more. For Bucky's geodesics have been discovered in the nanocosm.

DICK AND THE BUCKYBALLS

It turns out that what Buckminster Fuller did with artificial spheres like cannonballs, nature had already done with those sticky natural spheres called carbon atoms. Twenty years ago Dr. Richard Smalley of Rice University in Houston, Texas discovered that soot atoms can spontaneously arrange themselves into structures just like the U.S. Expo Pavilion. A Renaissance man who knew something about structural engineering as well as chemical physics, Smalley called these newly uncovered carbon allotropes *buckminsterfullerenes*—since abbreviated to fullerenes or, more playfully, "buckyballs." A buckyball is the first new type of carbon nanostructure found since graphite was imaged by X-ray crystallography four decades ago.

Before Smalley's discovery, only three forms of carbon were known. There was amorphous carbon, whose atoms were jumbled up every which way—that's regular soot. There was graphite, whose microstructure comprised wide thin plates that slid easily over one another, making this allotrope an excellent dry lubricant. Since it smears easily, graphite is also used in pencils, where it's mistakenly called "lead." Finally, there was diamond, a rare and precious crystal perfectly transparent to visible and infrared light, and about ten times harder than anything else in existence. Chemically, however, there's no difference at all between diamond and soot. We'll return to this odd fact later.

A perfect diamond is a single macromolecule. It comprises countless repetitions of a single module: eight carbon atoms, covalently bonded one to another in a strong, rigid, stable cube. This eight-atom module also links up with its fellows using covalent bonds. Diamond's unique physical and chemical properties make it suitable for many things besides adorning human hands. Industrial-quality diamonds, though inferior in size and quality to gemstones, slice easily through the hardest nondiamond materials. And when the U.S. National Aeronautics and Space Administration (NASA) dispatched a pilotless probe to land on Venus, a flawless half-ounce diamond went along for the ride as a window for one of the lander's sensors.

You want to know how far global warming can go? The surface of Venus is more corrosive than most acids and hotter than a self-cleaning oven. A polished diamond will withstand Venus's ghastly surface conditions, yet still transmit infrared (IR) photons to an infrared spectrometer kept inside the craft. Incidentally, here's a fun footnote for anyone who's dealt with government bureaucracies. When the Venus-lander diamond was imported into the USA for insertion into the planetary probe, the U.S. Customs Service classified the diamond as a "gemstone for export" and slapped a 44 percent duty on it. Customs officials relented only when NASA officials signed a sworn statement that the diamond was stuck forever 26 million miles away and would not show up later as a returned import.

As fascinating as diamonds are, they pale when compared to buckyballs. The buckyball molecule looks like a soccer ball. It is sometimes abbreviated C_{60}, since it comprises exactly five dozen covalently bonded carbon atoms in a geodesic nanosphere thirty angstroms in diameter. (An angstrom is a unit named in honor of the nineteenth-century Danish physicist Anders Jonas Ångström. It's one ten-billionth of a meter—i.e., a tenth of a nanometer.) In fact, one of the newest and most successful of the private firms springing up to exploit buckyballs' bizarre properties is a company called C Sixty Inc.

Like its big brother the U.S. Expo Pavilion, which is six billion times its diameter and 200,000,000,000,000,000,000,000,000,000,000 times its volume, the natural molecule called a buckyball is a perfect geodesic sphere. Also like its brother, and unlike its distant cousin the benzene ring, the buckyball is strong and rigid and does not flop around. Remember pharmaceutical design? If a conventional benzene ring changes shape, it can interfere with a drug's link-up to its intended attachment point, usually a molecular receptor on the outer surface of a human cell. Might the buckyball prove a better point of departure than the benzene ring for various synthetic drugs? Might it lead to treatments with greater efficacy, more consistent effects, and a higher, stabler, longer-lasting retention of medicinal properties? A lot of start-ups and giant biopharm multinationals think so, and they are betting millions of dollars on their prediction.

Mind you, there are contrarians. Dr. Neil Branda, for instance, will have nothing to do with buckyballs. Branda, an assistant professor of chemistry at Simon Fraser University, cites late-2002 research from Rensselaer Polytechnic Institute that shows how fullerenes can ignite with a *pop!* when subjected to a pulse of light in the presence of oxygen. This leads Branda to view both buckyballs and "buckytubes" (single-walled tube fullerenes) as unstable and potentially explosive. Of course Branda, who favors using viral protein coats, may not have paid enough attention to the troubles that gene therapy (GT) had in the mid-1990s with gutted or attenuated viruses used as treatment vectors. While Branda restricts his viral drug-delivery vehicles to viruses that harm only plants, viruses as a group are tricky things. At the very least, Branda's team could break their brains trying to visualize their chosen virus's exterior protein coat well enough to find stable attachment points for pharmaceutically active molecules. In the middle case, the bacteriophage could reconstitute its scooped-out guts, or otherwise mutate to a form that interferes with the attached medicine. At worst—this is admittedly a long shot—the viral protein may halt drug function entirely.

I agree with Branda that plant bacteriophages offer no likelihood of a doomsday scenario of direct human pathology. And even if drugs based on buckyballs don't explode on exposure to light, they may degrade via subsequent oxidation. But, as always, time will show who's right.

The caution of scientists such as Neil Branda hasn't prevented new companies from springing up to explore buckyball-based medications.

C Sixty had attracted $6.5 million in capital by year-end 2001, a mere twenty months after its incorporation. The company has five drugs under development, all based on buckyballs. The C Sixty drug nearest to acceptance by the U.S. Food and Drug Administration (FDA), the unofficial world authority on new-medicine regulation, blocks an HIV enzyme called a protease. This C Sixty drug may attenuate the severity or delay the onset of AIDS, and could be on the market as early as 2005. C Sixty's product has one major advantage over current benzene-based protease inhibitors: Patients using it do not develop accelerated resistance to other drugs.

Another C Sixty product aims to become the first known treatment for a genetic condition called ALS (amyotrophic lateral sclerosis), a.k.a. Lou Gehrig's disease. C Sixty is also developing a sensing and diagnostic toolkit to help medical researchers around the world monitor pharmaceuticals containing fullerenes. Using the kit, scientists can track the movement and concentration of any drug based on buckyballs.

C Sixty also hopes to develop an anticancer treatment designed to home in on cancer cells, bind with them, and then (and only then) unleash a cell-killing poison in response to light. Today's chemotherapy is like carpet bombing, attacking all cells indiscriminately: the healthy as well as the cancerous. C Sixty hopes its buckyball drug will act like a smart bomb, killing only cancer cells.

As interesting as these individual products are, they merely illustrate C Sixty's real intellectual property (IP): a lockup of many of the key patents governing buckyball use in drugs. Because of this, C Sixty describes itself as a "platform company" that will derive most of its future revenues from licensing and partnering.

THE PETAFLOP MACHINE

As their name suggests, "buckytubes" are a cylindrical form of fullerenes. Topologically they are buckyballs unzipped at one or both poles, and are only 1 nm in diameter. Discovered by Dr. Sumio Iijima in 1991, buckytubes may have single, double, or multiple walls. Any of these forms can show odd properties.

Take electron flow. In the meso- and macroworlds, amorphous carbon is used as an electrical insulator, whereas graphite is a decent conductor. Graphite comprises the same carbon atoms, not jumbled every which way but neatly arranged in planes. Alexander Graham Bell used granules of graphite to convert the information acoustically encoded in speech into data encoded as pulses of electric current. So modulated, the data could be sent faster and farther than the loudest shout via the "far-sounder" or, if you prefer Greek, the telephone. Packed tightly beneath a stretched membrane, the graphite granules pass more current when the membrane, driven by sound waves, squashes down on them. Current flow lessens when the diaphragm is at neutral position, and drops still further when the diaphragm flexes up.

Buckytubes show similar form-based conductance variations. When the carbon atoms in a buckytube are aligned straight with the tube's long axis, the buckytube is an excellent conductor—as good as many metals. Now imagine a pair of nanohands gripping either end of the tube and twisting it. Torqued in this way, each string of covalent bonds describes a spiral. Somehow, this turns the twisted buckytube into a classically perfect semiconductor. Come on, nano, let's do the twist.

It's fascinating to look at the graphs produced by IBM's Nanoscale Science Department (available at the Big Blue basic-research website). Beyond a certain twist limit, conductivity levels off so suddenly that the curve looks like a flat-topped mesa. If you prefer kitchen metaphors, it's like a watermelon that's been cleanly sliced off six inches from one end. However you regard it, the conversion from conductor to semiconductor is complete and abrupt. Furthermore, the bandgap of the carbon nanotube—a measure of the information it can transmit or store—varies predictably with the amount of cylindrical torque. Don't like your carbon as a conductor? Crank it into something else. Bending or twisting a buckytube changes its electron flow, just as crimping a garden hose restricts its flow of water.

On top of everything else, buckytubes are so tiny and so exactly shaped that they make ideal tips for atomic-force microscopes. In that case, they function as the sharpest knife points in existence, close to the ideal.

All this opens the way to nanoscale computer hardware. In 2001, IBM Laboratories demonstrated the first nanoscale transistor based on carbon nanotubes. There seems to be no theoretical barrier to multiplying such a device by any given factor, from two to a trillion. All these "nanosistors" could be combined in a single chip. A trillion nanosistors would create a VLSI nanochip (VLSI standing for very large scale integration). The VLSI chip would be a macro-sized central processor, only one centimeter (0.4 inch) square but with more raw computing power than all the supercomputers currently in existence. The nanotech CPU would lie at the heart of a "petaflop" machine, which by definition would be able to perform 1,000,000,000,000,000 or more discrete logic-gate operations per second (*peta* meaning 10^{15}). Combine this capacity with advanced, massively parallel processing, an architecture based on biomimicry of the working human brain, and you could have the long-sought basis of artificial intelligence.

This is speculation, granted; but unlike Drexlerian sci-fi, it's a sound theory. Even using the crude, brute-force approaches now in use (i.e., digital processing and linear architecture), a petaflop machine could one day approximate elementary human skills in intuitive thought, face recognition, and adaptive learning. So far such skills have proven well beyond the most advanced computers based on conventional microelectronics.

IBM's buckytube transistor made headlines when it was demonstrated in 2001. The enthusiasm of media reports on the invention often missed the real news: that by every measure, IBM's carbon nanosistor outperformed its big brothers based on silicon.

"The small [size] is of course very important," said Dr. Phaedon Avouris, manager of nanoscience and nanotechnology for IBM's T. J. Watson Research Center. Still, he added, the size *per se* was "a little bit overhyped. It is really . . . performance we are after."

Intel Corporation has subsequently issued press releases that reconfirm its faith in silicon's continued dominance. But what else could Intel have said? Would the master of a clipper ship praise steam?

In IT-targeted nanoscience, more than any other subfield except wet nanotech, basic discoveries can lead swiftly to commercial applications. Thus several other groups around the world are already on the trail of workable nanosistors. One of the most advanced teams is led by Dr. Paul Alivisatos, a chemist at the University of California at Berkeley.

The Alivisatos group at UCB is primarily interested in solid crystals at the nanoscale. Because this is reputable science, the group recognizes what Drexler et al. do not: that material behavior varies wildly with dimension, and that we must go to the nanocosm not as conquerors to dictate but as acolytes to learn.

"Many fundamental properties of a crystal depend upon the solid being periodic over a particular length scale . . . in the nanometer regime," says the Alivisatos group in a statement posted on the Internet. In other words, everything hangs on a material's nanoscale consistency.

"By precisely controlling the size and surface of a nanocrystal," the statement goes on, "its properties can be 'tuned' . . . [and] new nanocrystal-based materials can in turn be created."

In fact, entirely new devices can be created, one of them being an advanced single-electron nanosistor. Dr. Alivisatos, working in mid-2002 with Dr. Paul McEuen of the Lawrence Berkeley National Laboratory and Dr. Hongkun Park of Harvard, sandwiched a single buckyball between solid-gold electrodes to produce a nanosistor even smaller than IBM's buckytube device. The Alivisatos nanosistor gates the movement of electrons one at a time. With a tiny charge applied to the buckyball (measured in nanovolts, appropriately enough), the lone electron tunnels through the 60-carbon molecule from one gold electrode to another. In IT jargon, that simple action constitutes a "flop," the ability of a circuit to distinguish, on command, between *on* and *off*. The flop is a basic requirement of digital computation.

It's too early to say what type of nanosistor a VLSI petaflop chip would incorporate. At publication, it's a neck-and-neck race between IBM and Alivisatos *et al.* But petaflop computation is only a matter of time. When it happens in five to seven years, we'll see some major miracles.

STRONGER THAN SPIDERWEBS

As many of my interviewees pointed out, the properties of every material that we use in the macroworld depend on that substance's characteristics on the nanoscale. The fullerenes are no exception to this rule, particularly in structural values such as stiffness, strength, and elasticity. Materials that have a properly tailored nanostructure promise to be so efficient in the macroworld, so airy-light for any given value of tensile and compressive strength, that they may revolutionize our visible structures. Carbon materials, more than any other, offer us the opportunity of building things taller, wider, stronger, and safer than anything we have attempted before.

The key to all this is the single-walled buckytube. A scientist from the Houston firm of Carbon Nanotechnologies Inc. (CNI) waxes almost poetic about soot of this type: "The special nature of carbon combines with the molecular perfection of buckytubes . . . to endow them with exceptionally high material properties such as electrical and thermal conductivity, strength, stiffness, and toughness. No other element in the periodic table bonds to itself in an extended network with the strength of the [covalent] carbon-carbon bond." Not bad for a smutty beginning! To me, the key words in the CNI aria are *strength, stiffness,* and *toughness.* Durable stuff, carbon. In buckytube form, it's twice as strong as spider-web silk, which in turn is 100 times stronger than steel. A carbon-fiber cable two millimeters in diameter—the width of a ballpoint-pen refill—could support twenty tons and not come close to its theoretical breaking strain. At 25 tons, it would still be loafing. This is because buckytubes' actual strength comes far closer to ideal values than other materials can manage. In theory, low-alloy structural steel should tolerate 2,500 tons per square inch before it stretches and snaps. Instead, it fails in tension at one percent of that load.

To fulfill soot's structural promise, nanoscientists who would be nanopreneurs must first grow buckytubes far longer than they have yet done. Carbon nanotubes must come out of the sample dish and be produced in lengths of a hundred yards or more before they can support the Brooklyn Bridge with cables the thickness of a pencil. Already buckytubes have been grown in the laboratory to 100 microns, or a hundred thousand times their 1-nm diameter. Once this has been increased a further hundred thousand times, then entirely new species of bridges, buildings, and aircraft will soar aloft with no visible means of support.

Innovations tend to do two things, which they accomplish in sequential order. First they transform individual objects that already exist. This process is generally predicted at the time an innovation is first made. But then, quite unexpectedly, the innovation creates entirely new classes of objects—seemingly out of thin air. Buckytube nanotechnology is on the brink of doing this for structural materials and, through them, for almost every substance in our lives. Untearable fabrics, hair-thin auto bodies, nonpiercable body armor for police forces and the military: All will soon appear. Design is about to go anorexic. By 2015, we'll look back on the year 2003 as the Age of Clunkiness, when engineers built ungainly structures with laughably weak materials.

LAUNCH LEVEL, PLEASE

I've mentioned an ABC of existing things that C_{60} technology may modify—armor, bridges, and cars. But these applications are only the beginning. Buckytubes will also let us build things we had hardly dreamed about until a few years ago. One of the most impressive of these inventions will be an elevator into space. In size, weight, cost, durability, and function, it will be a tower of superlatives.

The invention of the staircase is lost in prehistory. Probably the first creature that could walk learned that it could gain height if it broke an ascent into smaller, manageable bits. Think how easy it is to climb a flight of stairs or, for greater heights, to take an elevator. Then imagine what would be involved in changing contour lines if staircases and elevator shafts didn't exist. Perhaps we could ascend using some heavier-than-air craft or by balloon, but both these things work only in the atmosphere. To get a mere fifty miles above our planetary surface, let alone to the moon and beyond, we need rockets—propulsive devices that provide thrust and lift in the absence of air. In terms of launching satellites, this dependency on flash-bang fireworks puts us in the Neolithic Age.

Now imagine achieving low earth orbit by simply getting in an elevator and pressing PENTHOUSE. Imagine launching a satellite by winching it up a permanent, hundred-mile tower and firing it out sideways when you reach the top. These are some of the revolutionary innovations that carbon-fiber technology may provide in the longer term—that is, by 2030 or so. Slimmer and stronger bridges, however earthquake-proof and aesthetically elegant, are just incremental improvements. A literal "stairway to heaven" is something utterly new. Yet this is what nanotechnology is on track to accomplish in the present century. Fittingly, it will do so not with buckytubes alone; it will also use geodesic designs. From the nanoscale to the extreme macroscale, the stairway to heaven would rely on Bucky Fuller's designs.

How might this happen? Let's use some disciplined imagination.

A hundred-mile tower to launch satellites would have to be grounded in bedrock somewhere along the earth's equator. Bedrock, for solidity; the equator, because that's where the earth's rotational surface speed is highest. Launching satellites from higher latitudes requires more work.

The project would be so expensive that only one nation could afford it: the world's sole hyperpower, the United States. South America contains the closest equatorial sites to the continental USA; and Quito, Ecuador is an excellent candidate. It lies exactly at the 0° parallel and has a Second World supply of power, materials, and workforce, all upgradable to First World standards with relative ease. Politically, Quito is more stable than Sumatra, Borneo, or anywhere in Africa, the other dryland regions that the equator crosses. Quito also adds the geophysical advantages of mile-high elevation and a hard rock foundation, attributable to the young, strong, lofty Andes Mountains.

The central tower of the Fixed Space Facility (FSS), nicknamed simply the Tower, would be an open latticework of individual struts. These would be resistant to both tension and compression and could be assembled in a Fullerene geodesic pattern called a dymaxion. The struts would range in length from one yard to a thousand yards, with longer struts braced at intervals between their end-points. A standard strut would be a foot in diameter, giving it a cross-section of about 120 square inches. This strut would have a minimum breaking strength in tension of more than sixty million pounds, or 300,000 tons.

The Tower itself would be square in cross-section and one mile to a side. It would be guyed with yard-thick buckytube cables at hundred-yard intervals along its length. On their earth side, guy wires would terminate in immense, immobile ground anchors called deadmen. The Tower deadmen would be set half a mile into the Andean bedrock and laid out along an enormous spiral. This ground pattern would begin due east of the Tower, with the first deadman only five miles away. Other deadmen would follow at 10-mile intervals until the final ground anchor sat 250 miles due west of the Tower base. This immense spiral would guy the Tower in all directions, offsetting both the prevailing westerly winds and the opposing, whiplash effect of earth's west-to-east rotation. That rotational acceleration would affect the Tower like a 1,000-mph wind blowing constantly from the east.

The center of the Tower would house an elevator shaft. Initially, this might seem crude: an open framework of buckytube struts, scarcely distinguishable from the Tower's structural elements. But it would be all that was needed. One of the vertical elevator struts could easily be banded at one-inch intervals with a kind of horizontal candy-striping, also made from buckytubes. This striping would wind around the central, vertical strut in a vast continuous helix. If uncoiled, the striping's path length would be over 200 miles. Satellite payloads would use the helix like a worm gear, inching their way upward. At a human adult's fast walking speed, they could reach the top of the tower in about a day.

Early launch walkers could be powered by liquid hydrogen and liquid oxygen, like the Space Shuttle. Later models could be solar-powered so that energy costs would sink to nearly zero and satellites could be launched, barring amortization costs of the Tower's construction expense, for little money. The Tower would place few maintenance demands on its owners, since the buckytube allotrope could easily be shielded from undue oxidation or abrasion from windblown particles.

At the top of the Tower, the launch walker would aim its payload dead horizontal, gently loosen its grip, and ignite a small rocket. This would give the satellite what engineers call delta-V_h, the required gain in horizontal velocity to take it to orbital speed. Thanks to earth's rotation at the equator, the payload would start with a V_h of 1,000 mph and gain additional V_h throughout its upward travel. Thanks to the Tower's height, the rocket would need to supply minimum energy to put its payload in a low-earth equatorial orbit. (A larger rocket would be needed to insert a satellite into a polar orbit; in a north-south direction, planetary rotation cannot act as a slingshot and a larger launch-push is required.)

Before the walker crept back down the Tower, it would switch from worm-gear propulsion to free fall and return to earth by gravity alone. The walker's braking systems are necessary to prevent it from arriving at ground level doing several hundred miles an hour and injuring itself on impact. The brakes would not simply dissipate the energy they extract and radiate it into space as waste heat, as car brakes do. Instead, they would store the kineticized gravitational potential in onboard batteries that would then help power the walker through its next slow-launch ascent. Energetically, this system would work like a counterbalance, offsetting the energy demands of each rising load with the energy gains from a falling one. Net result: Cost per pound of payload would make it almost as easy to launch your own satellite in 2030 as it is to launch your own Internet radio station in 2003. If the first FSS Tower proved successful, paying back its $150 billion initial cost in less than ten years, then plans would quickly be readied to build a sister Tower of almost twice the height.

The laws of orbital mechanics dictate that the taller the Tower, the greater its tip velocity and the smaller the rocket needed to launch a given size of payload. The new Tower could contain a true space elevator: a spacious, pressurized, mobile room equipped with viewing windows and communications gear. This would lift scientists and middle-class tourists into space fifty or a hundred at a time. One of the highlights of their journey would be launching their own satellites when they reach the top—the twenty-first-century equivalent of tossing a bottled note into the sea.

But this time the ocean is space, whose extent is not fifteen thousand miles but fifteen billion light-years. The Tower would be our first staircase into that endless ocean.

I CAN SEE CLEARLY NOW

The Age of Clunkiness clings to its final bastion the way France hung onto Algeria. Yet slowly, reluctantly, the last great icon of needless technical inelegance is finally being forced out of offices and homes, into museums.

I'm not talking of the wooden-shelled, wind-up Edison telephone of 1911 here. Nor the two-ton, four-door Chrysler Airflow that came out in 1934. I'm talking about that great clunky CRT monitor perched on your desk, or your family room's 150-pound, 36-inch-screen tube TV. These are both dinosaurs, and their days are numbered.

The cathode ray tube, whose thermionic vacuum-tube-powered ancestor first appeared eighty years ago, still sits at the heart of most computer monitors and entertainment TVs. And it's used in modern high-resolution and ultra-high-resolution transmission electron microscopes (HRTEMs and UHRTEMs). Even projection TVs are saddled with this invention, which is wondrously complex and prone to error.

In physical terms, the CRT is a linear accelerator. It liberates electrons from surrounding matter, then uses magnetism to increase their speed until they smack into a screen. Since electrons have a negative charge, they respond to an electromotive force and can be jinked around by electromagnets.

In early CRTs, electrons were produced by heat. A resistance element would literally boil them off into the low pressure of a vacuum tube—hence the term *thermionic,* from the Greek terms for "heat" and "going." Today's CRTs use solid-state technology, but for the electrons involved the principles are the same: Strip 'em, isolate 'em, accelerate 'em, and bash 'em into a target plane. This plane, the CRT's screen, is filled with phosphors—substances which, when pumped by fast-moving electrons, immediately dump their unsought energy gain by re-radiating it as visible light.

Other electromagnets, grouped in a tight assembly called a quadrupole, herd the electrons together into a tight, steerable jet. This particle jet functions as a kind of invisible paintbrush. The quadrupole raster-scans the jet from left to right, then up slightly, then from right to left. So quickly does it jerk around the electron jet that it can paint an entire CRT screen not merely once a second—thereby hitting half a million pixels—but thirty times a second. That's fifteen million pixels per second, every second, forever; each one uniquely filled.

If the successive full-screen paintings presented to us are slightly different, we slow-thinking humans are fooled into seeing smooth motion. That's because our own brains chop time up into 40-millisecond chunks, showing us the world as a succession of snapshots, each of which lasts 1/25 second. (It's no coincidence that this interval is also the duration of an eye blink. Nature has engineered things so that our natural windshields are washed, rinsed, and squeegeed in the unnoticeable interval between our 40-ms glimpses of reality.)

The CRT system, while complex, has proven so effective that it has remained conceptually unchanged for nearly four generations. From lab demonstration in the 1920s to commercial demo at the 1939 New York World's Fair, from Sid Caesar and Carl Reiner in *Your Show of Shows* to the 1963 JFK funeral, the 1969 moon landing, and the 2002 World Cup, television has been based almost entirely on cathode rays. TV is, literally, the gun heard (and seen) around the world.

The first changes to this technical monopoly came in the 1970s, when engineers commercialized a little-known laboratory novelty called the liquid crystal. In an LC display (LCD), an electric field determines light-transmission properties, making pixels dark or light. When multiplied by several thousand, the net effect is a black-and-white screen capable of showing words, line drawings, or half-tone images. LCD technology got a major boost a decade later when consumers demanded thinner screens for their small computers. (Unless you're so fat that your lap is desk-sized, there's no room on it for a full-sized CRT.)

By the mid-1990s, most LCD screens had full color. But even with innovations such as supertwist illumination, these screens faded in strong sunlight and showed different colors when viewed at different angles. What was needed was a video display terminal (VDT) that was as clear, true, and consistent as a CRT, yet as light, thin, and portable as an LCD.

Carbon nanotubes (CNs) have come to the rescue. Among their other talents, CNs can act as nanometer-diameter accelerators, briskly whisking electrons from rest to speeds that are high enough to excite phosphors on a video display screen. Thanks to their nanoscale properties, buckytubes can achieve threshold electron velocity in a length of half an inch or less instead of a standard CRT's 10–20 inches. In effect, a buckytube VDT is a

CRT with a highly compressed cathode-ray source. This permits a VDT that has all the advantages of a cathode tube, but that's thinner than your average ham sandwich.

Applied Nanotechnologies Inc. is located in Chapel Hill, North Carolina, two miles away from the University of North Carolina nanomanipulator. The company, a UNC spinoff, is one of the first nanobusinesses in the world to go (forgive the sooty pun) into the black. ANI's carbon-nanotube cathodes have excellent stability at current densities above one ampere per square centimeter, and they have achieved peak currents of 30 milliamperes. ANI's buckytube approach even works for the extremely energetic radiation known as X-rays. Here the thermionic CRT, unchanged in principle since Dr. Roentgen discovered X-rays a hundred years ago, is replaced by a matrix of buckytubes. The result is light, portable, fine-resolution X-ray technology that's still powerful enough to image the human body. ANI's nanotechnology is on the brink of putting X-ray diagnostics into every ambulance and first-aid kit. There seems no technical reason why a high-resolution medical X-ray scanner can't be smaller than a pack of cigarettes.

Of course, the application most likely to be the barn-burner for this buckytube video display technology, commercially bigger even than medical and health applications, is entertainment. Already Korea's Samsung and Gold Star, and the Japan-based multinationals NEC, Hitachi, and Sony, have R&D programs to explore and apply video displays based on carbon nanotubes.

There may be no practical limit to how thin a carbon nanotube (CN) screen can be. Already buckytubes can be routinely produced in lengths of a few microns, thinner than a coat of interior latex primer. If nanotubes could be persuaded to align themselves at right angles to a substrate—a feat that some experts think should not be overly difficult—then carbon nanotubes suspended in a paint matrix could function as a paintable video screen.

Take thirty seconds and consider the possibilities; I can think of several. Why paint any room more than once when you can change its color simply by turning a dial? Why shell out for another dedicated TV set when you can daub a permanent, functioning, state-of-the-art VDT on any surface you like? What's to stop you making your whole home into a virtual reality theater by having every square inch of its walls and ceiling project 3-D images?

All these things are not pipe dreams, like the Drexlerian molecular assembler. They are strong possibilities, probabilities even, that some of the largest firms in the world are spending tens of millions of dollars to develop. We can expect viable prototypes for the first of these fullerene-based inventions in under five years.

CHAPTER 9

SHIROTAE

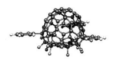

Spring nearly gone
And summer here,
Too soon as always!
For there the white shifts flutter
Against Perfume Bottle Hill.

— "The Empress Jito," English
translation from *The Little Treasury*
of One Hundred Poets, One Poem Each

THE WARM SWEET RAIN OF JAPAN

CALL ME A NIPPONOPHILE—a secret admirer of Japan, a worshiper
from afar. Something about the society, its delicate existence between the
extremes of stern duty and veiled emotion, appeals to me. And so in distant North America I have long read books, written my own clumsy *haiku*,
and pondered the Japanese soul.

One of my most cherished volumes is *The Little Treasury*, a medieval
courtier's compilation of "One Hundred Poets, One Poem Each." Modern
Japanese know it as a children's game. The hostess at a party reads the
first line of one of the poems, and the kid who correctly calls out the final
line is the winner. But if you're a *gai-jin* (foreign person) and an adult
besides, *The Little Treasury* is far more than a game. It's a peek into the
unseen Japan that lies behind today's nation of technical miracles and
robust democracy.

Shirotae. The word literally means "white mysteries." But in the Empress
Tenno's tongue-in-cheek poem, it represented the cotton underwear that

her people were sanitizing in the late spring air. As I undertook my research, the word *Shirotae* began to seem an ideal metaphor. The nanocosm is indeed a mystery, a layer that lies beneath our commonplace world and has lately been hung out fluttering before us.

While I've long admired classical Japanese culture, I had never been to Japan before I went in September 2002 to do face-to-face interviews for this book. For want of anything else, I decided to pay heed to commonsense advice. Some of it was excellent, given by friends and colleagues who'd been frequently to Japan. Take off your coat before you ring a doorbell; otherwise you're implying that you don't expect a welcome, which insults your host. When you go to an interview, bring some small gift as a token of respect. These tips proved excellent counsel, and I followed them to the letter. They gained me access to a Japan I would never have witnessed if my hosts had taken me for another blunt, blundering Westerner.

But then there was the other stuff. Dress formally, people said; even the garbage men wear a tie. Japanese are cold and distant, obsessed only with business. The entire archipelago is so overcrowded that there's hardly any such thing as countryside. The place is five times as expensive as Vancouver, so clear your credit card and take along a thousand dollars a day in cash for food and contingencies. You'll have to entertain; and when you do, count on dropping at least two hundred dollars per person per evening. Western news reportage of Japan's recent troubles is exaggerated; the Japanese themselves are full of pride and confidence in themselves, their culture, and their economy. And finally, Tokyo is dirty beyond belief. Not only is there grime everywhere, the air is so foul that you should spend all the time you can indoors. Never exercise there unless you're on a health-club machine.

I found every one of these clichés dead wrong. For one thing, while Tokyo is at the latitude of Los Angeles, it has one major thing that L.A. doesn't: water. The countryside from Narita Airport all the way into town is lush and green, with reeds growing in floodways and countless little fields of brown-topped rice. And it is a genuine countryside. There are vistas and expanses, long views with hardly a person visible. In the shallows stands the occasional patient crane. Granted, the signs of human habitation are always there—bridges, roadways, power lines, canals. But this is to be expected, for Greater Tokyo has a population above 20 million wrapped around Tokyo Bay. Oddly, I didn't find this enormous concentration of people, autos, and machinery intimidating. I grew up in a steel town, and Tokyo is identical, if larger. It's like a 10 × 10 grid of my hometown: a hundred Hamilton Ontarios, each of them holding a quarter-million people—a box of cities. But the *texture* of the two places, half a world apart, is a perfect match. Whether you're in the middle of Lake Superior or the Pacific, the view's the same. It doesn't matter that there's a lot more stuff beyond your immediate circle of sight.

In Tokyo, the view is Industrial City. Warehouses, fabrication shops, streets and highways, elevated rail links, apartment buildings, power plants. In from the countryside, the cranes are metal. Tokyo wasn't pretty, exactly, at least not in its approaches; but it was familiar. I have always had a covert, guilty love for the rough-edged places where people tear their fingernails and do hard work, the industrial sites that are a country's bones and biceps.

Yet even in this Big Smoke, I found a surprising amount of green. Antinoise barriers along freeways are engulfed by creeping vegetation soon after they're installed. The whole place is a kind of temperate rainforest; and each time it falls, the rain washes away the grime. At the time of my visit, Tokyo had just had heavy rainfall after twenty-one dry days, but that was enough to lay the dust and sluice the roadways. The rain also clears Tokyo's air. The morning after my arrival I stuck my nose out the hotel window and thought, This isn't so bad! And it wasn't. I held my run to three miles just in case, but although I'm an asthmatic my old lungs fared well throughout. And that was at morning rush hour, a kilometer from the Imperial Palace—smack downtown. Again, the key seemed to be the rain. In Canada, even in summer, the rain is chill, and slogging through it is unpleasant. The rain I ran through in Tokyo was refreshing, relaxing, and blood-warm—the gentle rain of Japanese autumn.

If a run in Tokyo is easy on the lungs, however, it's tough on the knees. I couldn't take a straight line; I was constantly braking to a halt and making tight right-hand turns to check out something I'd glimpsed. Tokyo is a place defined less by its public spaces than by its niches and alleyways; the city shows its real character on the nanoscale, so to speak. The big avenues at first seem sterile and tedious, but the instant you glance sideways there's something new, tucked away off the main drag. A Buddhist temple, rising imposingly from a cobbled square yet almost invisible from the sidewalk. A seafood restaurant, shuttered till dinnertime and the size of an average bedroom in North America. Storefronts selling pop and beer, T-shirts and toothbrushes. Nobody's taught them the mantra of location, location, location: They are where they are, and they know their clientele will find them.

My hotel is called the Hilltop. That's for us Anglophones; in Japanese, it's the Yamanoue. It's early postwar, with some later renovations, and once served as staff headquarters for the Allied occupation forces. It's near everything. Ochanomizu, the nearest subway station, is just around the corner; once I'm there it's simple to get around. The galleries at Ueno-onshi Koen are three stations and one transfer away. Tokyo University, where I have some interviews, is two stops and no transfers. This is an immense relief to me, because I've had some sleepless nights before my trip wondering if I'd get so lost I'd have to stay here. But I find transportation is ridiculously easy. Automatic ticket dispensers have an English option, and if I'm hesitant I know enough Japanese, and almost everyone I accost knows enough English, to set me straight. I've had more trouble navigating in Boston than in Tokyo, though I suppose that's not fair—the local speech in Boston is harder to understand.

The area right next to the hotel is a total gas: varied, lively, and unpretentious. From two to six in the morning, the streets close up to sleep, but the rest of the day they go full blast. Every fifteen feet there's a different café, noodle house, sushi bar, or open-air book market. Much of the vitality here is due to youth, for this is a young person's city. A baby boom in the good times of the late 1970s created a demographic spike among the under-30 that gives much of Tokyo a vigorous, temporary feel, as if half the people here are about to move on. That feeling of transience doesn't bother the youngsters any more than it does me. The only security, said a Japanese poet, comes in a hut built for one day.

My next surprise comes at how low the prices are. There's been some deflationary price-softening lately, and the yen has sagged beside the dollar, so I never find the outrageous bills I've feared. Any number of clean back-alley restaurants serve fabulous food for what I'd pay at a restaurant in downtown Cleveland. The Hilltop charges 17,000 yen per night, about $140 U.S. And try as I might, I cannot entertain—everyone I interview buys me a drink and a meal. I'm treated like a minor celebrity. My age brings me respect. My profession brings me respect. I'm even average height, the first time in my life that's happened. It's unnerving at first; then I start to enjoy it. What a confidence builder! Reentering suburban existence, with wife and children treating hubby/dad in their usual way as a harmless, well-loved moron, creates a feeling that puzzles me until I recognize it. It's cold turkey—the abrupt termination of an addictive substance. Withdrawal of esteem.

Strangely, I find it a writer's dream not to know the Japanese language. Understand what's spoken, and you're swept away by a flood of nonessentials. From poems in translation I know that Japanese has a suppleness of expression embracing the most tender, indirect emotion and the most brutal physicality. But like all languages it spends most of its time in latency, *capable* of sublimity but usually called upon only to support workaday life. In my ignorance, I'm spared this quotidian trivia, the Japanese equivalent of *Whaddidya say? Nothin'. Whaddaya mean, nothin'? I mean nothin' nothin.'* Absent this, I live in the clear bright world of the nonverbal—sun, wind, and weather; smell of little restaurants up laneways; a striking statue of the Fasting Buddha; the soft curves of a bridge.

One thing puzzles me. There are no litter bins in the city—and also no litter. In a half-hour run I don't see so much as an empty pop can. Every second person is smoking, but there isn't a cigarette butt in sight. And then I see a smoker finish a cigarette, tamp the butt out gingerly on a lamp post, and *tuck it away in his cigarette package.* My God, no wonder there's no litter in Tokyo. They'd eat it if there were no other way to dispose of it.

Most of all, I like the tiny grace notes of the place. White gloves on the taxi drivers. Service staff who (gasp!) actually like to serve. The warmth, the acceptance, the open-hearted friendliness I meet everywhere is staggering. Back home doormen and janitors, waitresses and sales clerks, tend toward two expressions: half-veiled contempt and open contempt. Their disgust extends to everything—employers, customers, themselves. In Tokyo, it is possible to have what the West regards as a menial job and perform it with address and panache, giving exemplary service and receiving honor and respect. Tokyo has *professional* waiters, *professional* bus drivers: I've never seen anything like it in my life. The entire world should work like this.

I dwell on the nonscientific aspects of this society because I've found over decades of writing about science in business, and business in science, that both subcultures are subsets of the overall culture that supports them. There are ways of doing business that are demonstrably Swiss, Australian, American, or Japanese. Nanotechnology originating in Zurich is as exact as a wristwatch, with a watch's conservative, instantly recognizable uses and high profit margins. Nanotech from Boston, Dallas, or San José is packaged cowboy-style, with applications that are outlandishly imaginative; seat-of-the-pants management and financing; and an all-or-nothing, pedal-to-the-metal commitment. Nanoscience or nanotechnology from Tsukuba or Tokyo, Hokkaido or Honshu, should by extension have a form that's uniquely Japanese. I'm here to discover what that is.

My body, bless it, has unexpectedly adapted to a sixteen-hour time difference overnight. I rise at six and run past the enormous stone drywalls that flank the Shogunate fortifications of the Outer Citadel. The great trees drip warm rain. I shower, dress, repack, check out, and head to Tokyo Station to catch a bus. For the next three days I'll live and work in Science City.

HIGHWAY-BUSU TO TSUKUBA

Japanese public transit is exemplary. In most North American cities, taking a bus is like cleaning your basement—it's necessary once a year or so, but you allow a full day for it and you never do it in good clothes. In Japan, things are different. Tokyo trains and subway cars run to the split-second and are as clean as a whistle. So are the Japanese highway buses, the justly famous *highway-busu*. They're modern, powerful, and gleaming inside and out. In fact, washing vehicles seems to be a national obsession. In all my time in Japan I saw only one dirty car—and it was driven by a big-nosed foreigner like myself.

I get down from the *highway-busu* where my directions indicate. As the bus pulls away, its engine sound is drowned by an even more deafening roar from the nearby trees. It's autumn, typhoon season, and the cicadas are waking for their six-week party. They're so loud I would have to shout to make myself heard.

For today's journey, I have received a map and written instructions from Dr. Tsunenori Sakamoto via e-mail. Dr. Sakamoto is deputy director of international affairs for AIST, Japan's National Institute of Advanced Industrial Science and Technology. In Japanese nanotechnology, AIST is largely where the action is.

With some exceptions, whom I'll talk about later, Japanese university researchers are modeled on the traditional English ideal: that is, they are strictly curiosity-driven. While their work may be brilliantly original, the link between it and commercial products is usually indirect and often nonexistent. For their part, the large Japanese corporations play new-product development close to the chest and guard their IP portfolios jealously. Besides, over the last decade of turbulent economic times many big companies that I had thought to have indefinitely deep pockets closed their in-house labs and outsourced product R&D to North American knowledge firms. Because of all this, AIST has emerged as Japanese nanotech's center of gravity.

Tsukuba, Japan, is in Ibaraki Prefecture, eighty kilometers north of Tokyo. Twenty-five years ago the place was a sleepy agricultural village. Then, starting in the 1970s, the Japanese government dropped an enormous series of R&D institutes onto the place, instantly quintupling its size. As well as quantitatively, Tsukuba was changed qualitatively. It's now a true Science City, with square kilometers of laboratories, workshops, and offices. Being put up all at once, the new city had the luxury of thorough planning. Buildings are not tossed on the landscape any which way, but are sited carefully around rushy, carp-filled ponds and surrounded by 200 kilometers of tree-shaded walkways and cycle paths.

Dr. Sakamoto meets me in Tsukuba Central 2, the main administration tower, and ushers me to his office. I'm grateful for the strong coffee that his assistant brings. We do the Japanese acculturative introduction dance—neither of us certain whether to shake hands, nod, bow, or all three—and sit to talk. I haven't mastered the subtleties of the bow. I have no clue when to start or finish, how many genuflections to make, how low to bend, or where to put my arms. Half the time I hear my heels snick: Colonel Klink without the monocle. My bows must be the cultural equivalent of a damp, limp handshake.

Dr. Sakamoto has been a godsend. I've been able to identify nanotech hotspots throughout Europe and North America using the Internet. I review work, identify key scientists, locate their home page URLs, and derive e-mail addresses sitting in my office. In many cases I conduct interviews electronically, saving myself weeks of time and a fortune in airfare. But Japan proved impossible to crack this way. As it did twelve hundred years ago, the place works via human contact. If you know someone who knows someone, you're in the door. If you don't, you're out of luck.

What broke things open for me was meeting Dr. Yasutaka Tanaka, a professor of chemical physics, at a friend's house during a Vancouver barbecue. I explained what I was doing and how difficult it had proven to make initial contacts. The scientists I did track down ignored my e-mails, or else sent back coldly polite kiss-offs pleading faculty review meetings, grant application pressures, the wrong phase of the moon, etc. Yasu listened to my plaints, then slipped away quietly. He came back in five minutes with a Mac Titanium that held his global address file. I stood slack-jawed as he rattled off contacts, all of whom were directly known to him. This was treasure. He put his finger on one address-book line: *Dr. Masumi Asakawa. Senior Research Scientist, Nanoarchitectonics Research Center, Tsukuba Central 5, 1-1-1 Higashi.*

"Masumi's a buddy of mine," Yasu said. "I'll e-mail him tonight and tell him to expect your message." And that was it. *Boom,* the door of the vault flew off. Like many senior scientists at Tsukuba, Yasu had done a foreign postdoc that had taught him excellent spoken and written English. And Yasu was friends with Masumi, and Masumi with Dr. Sakamoto, and Dr Sakamoto with . . .

When I arrived at Tsukuba, I had several long days of interviews lined up. Masumi, who was my host and guide on day two, added many others to the list for me to talk to. He was so well regarded as a scientist, and so well liked as a friend, that he could call whomever he chose and instantly set up an appointment. He, and before him Yasu Tanaka, were the keys to all that I learned in Japan.

BACK TO THE historical present. Dr. Tsunenori Sakamoto is brisk, efficient, and dressed in a crisp, putty-colored suit. We swap cards. I like the Japanese way of tendering things, whether business cards or cups of coffee: with both hands, fully facing, and a bow. The Tao teaches that the Profound is also the Subtle, and this tiny gesture says volumes about respecting others and oneself. We in the West thrust out our business cards like switchblades; or worse, we flick them onto tabletops as if tossing chump change. Making an exchange of cards a ceremony increases both participants' honor. It's an interview's perfect start.

Having so instructed me in manners, Dr. Sakamoto proceeds to astonish me; his first words demolish another slice of my conventional wisdom about Japan.

"Our scientific and technical achievements at AIST are not insubstantial," he says. "We have a lot of patents. But our conversion of this patent portfolio into revenue must be described as miserable. It handicaps us, this inability to get money from industry. We know so much, and earn so little! We must do better at it. At the moment, the government supplies 99 percent of our capital budget and operating fund. We hope to change this, starting very soon."

Dr. Sakamoto outlines the depth of the problem in disturbing terms. "Revenue from semiconductor chips peaked in 1987, when Japan had 50 percent of the world market and the U.S. had 35 percent or so," he tells me. "But since then, the U.S. has had a steady gain and Japan a steady decline. Our latest figures show the U.S. with 57 percent of the world market, more than we had fifteen years ago. Japan's share is down to 29 percent and, apparently, still falling. This is a severe problem that we face."

Dr. Sakamoto doesn't say so, but I strongly suspect that the sales curves he's showing me are capital-investment curves, shifted five years to the right. You spend money to make money: That's a truism. Even a cash cow needs hay. I ask him about this, and he smiles wryly.

"What is there to say? In the 1970s, when money was much more scarce, Japan somehow found nearly $600 million U.S. to capitalize its semiconductor R&D. That subsequently paid off twenty times over. But we got complacent, and did not keep up our investment. And so our revenues fell."

What's the solution? "It is partly strategic. To this end, AIST has been restructured. Before April 2001 we comprised fifteen loosely affiliated, semi-independent institutes that were all part of MITI, the Ministry of International Trade and Industry. Now we are at arm's length from a new government organization—METI, the Ministry of Economy, Trade, and Industry. And we are more tightly organized within ourselves. AIST is now a single institute." What do you hope to do with this reorganization? "Get our various researchers talking more with one another. Get them also talking more, far more, with industry. And with our universities."

AIST, Dr. Sakamoto explains, has 2,500 permanent scientists, an additional 2,200 visiting scientists, and a lean administration of less than 700. As has happened at R&D institutes all over the world, technical support staff has been pared. Now most AIST scientists must purchase and set up their own experimental devices. While this cuts into actual research time, it does save millions of dollars on technicians' salaries. Whether or not it wastes more money in the form of lost research time is unclear.

What about the universities? I ask. Don't they have ten times more scientists than AIST? Dr. Sakamoto frowns. "For many years our universities have had a nickname. We call them 'the coffins of the brain.' They collaborate with other universities, but hardly at all with colleagues in Japan; and as for industry . . . .

"Ah, well. I believe your expression is 'ivory tower,' is it not?"

While the new AIST-METI structure is vital, Dr. Sakamoto says, it will not succeed in regalvanizing Japanese industry all by itself. "The problem is not with the infrastructure so much as it is with the professors. I will give you an example: My own area is semiconductors. When it grew apparent a few years ago that world semiconductor sales were dropping, a call went out to the universities to tender new ideas. Back came the answer: 'It is too expensive for us to do anything in this area! Let the companies do it: We have nothing to offer.' Nor did they. There were no new ideas, and thus our industry fell behind."

To address this situation, the Japanese government has established nanoscience and nanotechnology as a flag project around which all aspects of the nation can rally—government, industry, universities, and AIST. The country, in other words, is betting its future on nanotech.

"The old days, the glory days of 'Japan Incorporated' have passed," Dr. Sakamoto concludes. "It is not impossible to get them back, but it will take work. Having lacked a national focus for so long, each private company has gone off in its own direction. There has been no consensus about what goals the whole nation should work toward. I believe that nanotech may give us such a focus."

LINGUA ANGLICA AND THE SLEEPING COMPUTER

Dr. Yoshishige Suzuki is young and ebullient. He's an expert in a kind of electronics that is not, by traditional definition, a form of electronics at all.

Like all fundamental particles, the electron has a specific set of characteristics that seem invariant: These properties define the thing. Electron mass, for example, is one-eighteen-hundredth the mass of the proton, a nuclear particle with positive charge. Electron charge is opposite to proton charge, so that the two particles strongly attract. Bound to protons in this way, electrons constitute every known atom.

Electrons have another property you don't much hear about: spin. This term may be a metaphor, as "color" and "charm" are for the sub-sub-atomic particles called quarks. Then again, maybe not. Electrons really do seem to whirl about an internal axis, which gives them a derivative quantity called spin angular momentum. The direction of its spin axis reveals the polarity of an electron's magnetic field.

Classical electronics, which is to say everything involving electricity, harnesses only the charge of the electron. But Yoshishige Suzuki is interested in harnessing electron spin, which involves a brand-new discipline known as spintronics. Dr. Suzuki is group leader of the Spintronics Group at AIST.

"The standard semiconductor," Yoshi tells me, "takes no notice whatsoever of electron spin. But science has linked spin to charge, making spin a useful property to investigate for new products." Is his research curiosity-based? "Originally, yes. But now AIST has a mandate to pursue commercial applications." New apps are emerging, Yoshi says, because of other research in nanoscience. "Below 60 nanometers or so, we can create what we call 'single-domain particles.' These are the smallest possible permanent magnets, other than the electron itself. We can control the spin orientation of these particles—that is, the direction in which their spin axes point. This makes possible a kind of spin-transistor at the nanometer scale. We hope this will lead to a high-speed digital logic gate, a device that flops between detectable states in less than a nanosecond. If we can [produce] large quantities of such a thing, we will have MRAM, or erasable-rewritable memory based on magnetism."

How hard is it to make a lot of these spin-based nanosistors with good quality control? "Hard, but not impossible. We are working on just such a project, along with Sony, NEC, and Toshiba. In the USA, Motorola and IBM are also pursuing MRAM."

Applications? Yoshi gestures at his notebook computer. It seems state-of-the-art to me, but Yoshi dismisses it as Neolithic. "This thing uses too much power. It is always on, which drains its battery. People are always wanting more powerful batteries; but the problem is in the computer's power drain, not a battery's ability to supply it. With MRAM, we could make an instant-on computer, one that boots up and loads programs in a millisecond or less. The normal state of such a computer would be off. It would sleep most of the time." Explain, I say, and Yoshi obliges.

"Say you are typing. The always-off computer would wake up only to receive and process each 25-millisecond keystroke. Then it would shut down completely for 75 milliseconds, until your next keystroke. Thus it would not need any power 75 percent of the time. A standard lithium-ion battery could power such a computer for up to a day without recharging, as opposed to two hours or so today."

At lunchtime I wander into the building's cafeteria, where I find faces and accents from around the world. English is as common in this room as Japanese, which I suppose is only logical. If you're from Benin and I'm from Mumbai, the tongue we're likely to share is English. *Lingua franca* has become *Lingua anglica*—the means of choice for scientific interaction. Esperanto, move over: There's a world language now, and it's English. Through sheer dumb luck, I've become an international author.

LAMINAR MAN

My guide today is Kenji Kawai, Dr. Sakamoto's assistant in international affairs. He's a pleasant, helpful young man who doesn't quite know what I'm doing, but has evidently been told I'm to be handled tactfully. I'm what the National Research Council once called a "visiting fireman"—a VIP.

Thank God for air conditioning. Japan, like the U.S. Sun Belt, entered the First World largely because of its newfound, technology-based ability to temper heat and lower humidity. Without air conditioning, my brain would be good for nothing but hanging around languidly in silks. And this is only September: I can dimly imagine trying to work here in July.

Kenji delivers me to a real force, Dr. Akira Yabe. Dr. Yabe is that rarity, a professional engineer with a doctorate. Such people are messianic. They combine the drive of the engineer, the rock-solid faith that an imperfect world needs them to give it order, with the talent to do precisely that. Dr. Yabe is director of AIST's Research Center for Advanced Manufacturing on the Nanoscale. He is also professor of engineering mechanics and systems at the University of Tsukuba, adjunct professor of the Co-Operative Graduate School of the Science University of Tokyo, and adjunct professor at the Kanazawa Institute of Technology.

Dr. Yabe rises from his desk to greet me. I feel as I did when I was sailing off Maui and witnessed a breaching humpback whale: This is a very impressive entity. Dr. Yabe's specialty is energy conservation. Not the theory of it, but the practicalities—how to study it, how to do it. His major at university was heat-pump systems. Right now he's looking at hot and cold water.

"Hot water!" he says, pumping my hand. "Such a waste! So much of it is flushed away after use—baths, laundry, and so on. If we could recover even part of that waste energy, we would save vast quantities of energy. Money, too. But such problems are legion. It's not the problems that are scarce, but the ideas to deal with them. Many technical needs! Few technical seeds!" He grins. I sense this is a line he's used before. Of course that doesn't make it any less true. From hot water, he turns to cold.

"I have been investigating a natural protein that acts as an antifreeze," Dr. Yabe tells me, plumping himself back down behind his desk. "This is found in certain fish and frogs in arctic regions. It keeps water from forming its usual solid crystal form when conditions get sufficiently cold.

"The protein is mostly hydrophobic; that is, it repels water. It is 'water-hating.' This protein is arranged as a long, linear chain of amino acids. Yes, yes, I know all proteins are configured in this way. But this protein *stays* linear. It is like one long arrow. The only places on this protein that bond water, that are what we call hydrophilic or 'water-loving,' are its ends. So what do you think happens?" I say, truthfully, that I have no idea. "It influences the development of ice crystals!" Yabe shouts triumphantly. "It makes them long and thin, like needles. Like the protein itself. Now what is a characteristic of such modified ice crystals? They do not form large, solid blocks. They make the water into something like thin, pumpable slush. A slurry, is the term.

"I am investigating the synthesis of this natural protein in large quantities. It could be added to the chilled water that we pump through buildings from a central compressor, for distribution to local air-conditioning units. In such cases it would form a solution that is still pumpable even at temperatures below the normal freezing point of water. The delta-T would be greater, yes?" I'm sure it would, I say. Just tell me what a delta-T is. "The difference, the temperature difference! Greek letter 'delta' for English term 'difference.' Delta-T, you see. T for temperature.

"So! The water we then pump is so cold that it can absorb more heat. Yes? Another way of looking at this, the mathematical equivalent: The water has more *cold* that it can give up to its surroundings. So we have automatically a more efficient air-conditioning system."

Okay, I say, fine. But this solution, suspension, whatever, it's more viscose. You couldn't pump it as fast.

Yabe spreads his hands, laughs hugely. "So what! You would not *need* to pump it as fast, would you? You would not *want* to pump it as fast. You would require the fluid to stay longer in the pipe so that this solution could extract more ambient heat from your room. The slower pumping speed would not be a problem at all."

Where does the nanotechnology come in here? I ask. Yabe lights up further, if such a thing is possible. "Hah! Nanotech means we *tinker*, does it not? We are not satisfied with what we find naturally, however good it may appear. So we tinker with this natural antifreeze protein. Biomimicry, not biotheft! We do not simply take what we find. Polyvinyl alcohol is what we call this substance. It is not toxic to people, to pets, or to the environment, unlike other heat-transfer fluids—for example, the ethylene glycol in your car's cooling system, which is a vile molecule and terribly poisonous. It tastes sweet, so your cat laps it up, and then—*Dead!* We avoid this."

Dr. Yabe concentrates on manufacturing, he explains, because it is the key to a robust economy. There is an unending need for new products that can be manufactured and sold. Demand for such products will never dry up.

"We want to use nanotechnology," he says, "not so much to create nanoscale devices as to create macroscale objects. To make sellable manufactured goods, with sellable new properties." The Drexlerian approach, he says, is unworkable. It would take too long to make full-sized consumer goods by building them one atom at a time. "Yet at the same time, we do as nature does. We combine the functions of processing and assembly. In traditional manufacturing, these functions have been totally separate; now they must be one. Otherwise we shall be left with bins full of parts that never, however beautifully designed or cunningly made, never act in concert as a single system, a single thing." So you're both an experimentalist and a theoretician and an engineer? I ask. Big grin. "*Hai!*" That, I remember, means *Yes!*

Dr. Yabe has one more surprise in store for me. "You know the conventional theory," he says, "that a smooth, solid surface minimizes the flow of fluid over top of it?" I didn't, but I'll take his word for it. "Well!" he chuckles, rubbing his hands. "Look what we have done!

"See this surface. It is silicon oxide, much like glass." I take the small sample he gives me and turn it over in my hand. It reflects back the white window light in many colors. "It's a diffraction grating," I say.

Dr. Yabe beams at me; I've answered the professor's question properly. "It is! Now see what happens when I shine coherent light on it." Yabe zaps the sample with his laser pointer. As I saw when Angela Belcher lectured in California, a regular geometric pattern springs up on the wall.

"Diffraction pattern!" Yabe exults. "We have incised nanoscale trenches in this otherwise smooth surface. Lines four microns apart and six hundred nanometers deep. Now what do you suppose happens when water flows over this surface?" I shrug and make the logical inference: Turbulence and laminar effects will hamper the fluid's flow, increasing its coefficient of moving friction.

Yabe slaps his thighs in glee.

"No, no, no! F-sub-C falls by a factor of two! Now what do you think of that?" I blink at him: Say *what?* "Water moves across this scored surface twice as easily as across an unscored one," he explains. Great. Why? "The moving fluid traps tiny air bubbles at the bottom of the trenches. It then flows with almost no resistance across these tiny trapped pillows of air." Applications? "Lower energy drain on almost every pump in existence in the world," Yabe says, dead serious this time. "Think of the energy, think of the *dollars*, that will save."

LADIES AND GENTLEMEN, BRACE YOURSELVES

You're on the subway, going to work. It's crowded in the worst way, enough so there's no handhold available, but not enough to jam you in place whatever happens. The train shrills around a tight bend and at the same time decelerates sharply to stop at a station. Not only do you, an experienced rider, stay on your feet; you don't even skip a word of your newspaper article. In fact, you hardly know what you've done until it's pointed out to you. Amazing thing, the body. Practically runs itself.

If Dr. Yoshio Akimune has his way, skyscrapers will one day do what subway riders do. They will sense perturbations in their physical environment and adjust to them so quickly and effectively that people in them will scarcely know what they've been spared in the form of shaking, swaying, or collapse.

Dr. Akimune is deputy director of the Smart Structure Research Center (SSRC) at AIST. As the center's name implies, Akimune and his colleagues are working on a way to make structures act as if they have intelligence—reacting instantly to minimize or eliminate the mechanical harm that too often makes buildings crack or crumble.

Japan is one of the most seismically active regions on earth, so Dr. Akimune's smart structures are largely designed to resist the effects of earthquakes. But it doesn't require too great a leap of imagination to see that a smart structure would also be strongly resistant to hurricanes, typhoons, or acts of terrorism. By definition, a smart structure is intelligently stable. Even when shaken, it lands on its feet.

The SSRC is nominally directed by Professor Fu-Kuo Chang, a Taiwanese scientist who also has a cross-appointment at Leyland Stanford University in California. The advantage here for AIST is having one of the world's leading smart-structure scientists as its research leader and project designer. The disadvantages, I infer as I keep my ears open, include Dr. Chang's *modus operandi* as an absentee director, plus the difficulties that arise when a Japanese speaker and a Mandarin speaker try to communicate engineering subtleties in English as a second language. Despite the obstacles, the SSRC has been hot on the trail of making smart structures work. But I'll let Dr. Akimune do the talking.

"We are fortunate," he says, "because this program creates technology that is more easily transferable to the commercial sector than some basic research. This is why our organization is configured as a Center rather than as an Institute—institutes within AIST are directed more at a basic understanding of nature."

Smart structures, Dr. Akimune explains, constitute a special case of biomimicry: "We intend to duplicate the abilities of living organisms to achieve and maintain physical equilibrium. In place of brains, we use computers. In place of natural senses, we have dedicated sensors. These detect undesirable effects such as imbalance, vibration, translation [sideways displacement], or rotation. Finally, in place of muscles, we have devices called actuators. These convert input energy, in the form of electricity or acoustic wavelengths [sound], into mechanical energy." So your sensors pick up something happening to the building that you want to counter? "Yes, and they do so at a very early stage. The computer then determines where, when, and for how long to apply countering forces, and tells the actuators what to do.

"Of course this involves feedback loops, *neh?* During and after actuator operation, the sensors continue to tell the computer how effective its actions are. The computer will continue to adjust its actions to minimize damage." Okay, but where does the nanotechnology come in? Dr. Akimune smiles.

"We use nanoscience to understand, at the smallest possible scale, what happens to structures when they undergo stress. Then we use nanotechnology to engineer the components of a smart-building system. We do this by building sensors and actuators right into our building components." You mean these devices are installed at the optimum locations in the building? "No, it's far more elegant than that. We can build them right into the building materials. Structural beams and joists, for example, can incorporate sensors and actuators in their very microstructure, almost within their nanostructure. In addition, we can tailor the thermal-performance characteristics of our smart materials, or even their crystal structure at the nanoscale, to be exactly what we want. The idea material would be made 'transparent to vibration,' so to speak. This is our ultimate aim."

The embedded micro-electro-mechanical system (MEMS), says Dr. Akimune, could be built up in layers only a few molecules thick. "When an electric current is passed through them, they would move. It is called the piezoelectric effect." Isn't that what turned a phonograph needle's wiggles into electricity in the days of vinyl records? "Correct. But the effect works both ways. The phono cartridge turns mechanical motion into electrical current. Our actuators go the other way around, and convert electricity into movement."

Smart structures from smart materials: It's a good idea. It is, moreover, one that can be applied to structures other than buildings. When a high-speed train goes into a narrow tunnel, Dr. Akimune tells me, the airborne shock wave coming from its bow bounces off the tunnel walls and is reflected back to the body of the train. This can cause a great deal of noise and vibration. "It is uncomfortable for the passengers to be buffeted about in this way." The solution? Treat the train structure, which is often made like an aircraft using a technique called stressed-skin or monocoque, like a building. Use smart materials with custom-designed thermal and crystal properties. Embed sensors and actuators. Then let the *train itself* figure out what to do when the reflected bow wave comes calling. Elegant.

"One could do this for cars," Dr. Akimune says, "to reduce unpleasantness for passengers when driving with an open sunroof, or in a cabriolet [convertible] with the top down. Or at high speed. Though I am told that here marketing considerations come into play. A young man driving a car at high speed *wants* to feel speed, even at the cost of some discomfort." I smile: What about going the other way? Produce a smart car that exaggerates speed effects at lower thresholds, giving the "racing hit" at slower speeds but making roads, drivers, passengers, and pedestrians safer? Dr. Akimune sighs. "Again, marketing. You are dealing with people here, not structures. Structures are complicated, but they can still be understood. People are not so easily understood. They are more complex still."

WHEN KENJI KAWAI comes to take me away and register me in the AIST campus guesthouse, I'm beat. It's hot now, hazy and humid, and the cicadas are louder than a midnight rave at the neighbours'. *Sakura-kan,* the guesthouse, is named for something that has delighted the Japanese since ancient times: the cherry blossom. And indeed, the Tsukuba campus is filled with cherries. While the trees have long since shed their blossoms and are soon to lose their leaves, something of their magic remains in the house that's named for them. It's a three-story affair, which also houses meeting and planning rooms for international scientific conferences. Like most of the Tsukuba campus it's modern, brick on the outside and gleaming inside. Past the automatic sliding doors that are everywhere in Japan, there's a hush like that of a great library. A vending machine tucked away in a niche near the reception desk dispenses ice-cold Kirin beer for 290 yen a pint. Ah, civilization.

Behind the reception desk are a battalion of white-haired gentlemen who share the quiet pride of aristocrats. They are unfailingly helpful and tolerant of my broken Japanese, and I walk in terror of offending them. *"Cardu!"* one says, making a two-handed, bowing presentation of my identity card. Here's the system: You don't take the key with you when you go out. That would be difficult, as it is stapled to a lump of Lucite the size of a sledgehammer. Instead you leave the key at the reception desk with the white-haired *monsama* who's on duty and walk off unencumbered except by your small paper card. It's a hassle-free way to work things.

Out front there's a stable of well-used bicycles lined up in a perfect row, each cycle in its proper numbered slot. These are the loaners of Cherry Blossom House, used to expedite short-range transportation. I look at them with my head tilted to one side. I haven't been on a bike in years, but something about these ancient warhorses tempts me to be a madcap. They're battered and rusty, but the seats are adjustable and they seem mechanically sound. Perhaps if I changed into something more comfortable . . . .

I bow to the monsama, find my room, unpack my cases, and hang up my suits. Oh my, do I have suits. And I'd trade 'em all now for a decent pair of jeans. Japan, it seems, has suddenly become a lot more relaxed in the last few years. Dr. Sakamoto wears a suit, but he's an assistant director; the bench scientists don't bother. Most put on a tie on only for the monthly section meetings. As for the grad students, they are of a piece with their species everywhere: dressed down to the point where they'd make a bum blush.

By contrast, I've come loaded for bear. For interviews, I have two double-breasted, four-button units in three-season wool, one sharkskin and the other smoke-blue. It nurtures respect, like my advanced age and my assumed rank of foreign expert, but I wonder if I'm not a figure of fun to my casually outfitted interviewees. I'm like the pool-party guest in a tux: overdressed to the point of absurdity. And it's *hot*. Oh, is it hot. One good thing about my suits, I realize as I hang them up, is that they're self-ironing. In this hot, humid climate, no wrinkle lasts thirty seconds. Put it on my tombstone: *Best-dressed man in Tokyo.*

Downstairs to the gate-lords. *Bi-cy-cru? Hai!* No charge, sir; you are our honored guest. Take care, however, as it is almost dusk. The night is falling and the cicadas are shutting down.

Out I go. It's true what they say: You never forget how to ride a bicycle. Trigonometry, yes; algebra, geometry, the French and Latin prose. But not a bike. I flip a cam lever, twist and lift the padded seat, and swing my leg over the crossbar. *Move labouring out into the bourneless night.* No light on this thing, no reflectors, but maybe I can find a bike path. Hey, there's a bell! *Ting-ting!*

Damn, the gearing's high. Maybe if I twist the handle? Ouch! Yes! Nearly cost me my scrotum when the lower gear cut in and the pedals slipped, but this is much easier. Wind in my hair, haven't felt that since the do-gooders at Vancouver City Hall passed a helmet law ten years ago. I'm working hard, but with my slipstream's delicious breeze I'm cool for the first time in three days. Gear up again; out along the ring-road that orbits AIST Central. Whee!

Hey now, look at this. A narrow path through darkling woods, whilom the gray thrush singeth. *Whilom,* that's a word you don't hear much. A tight turn—

Unh-Unh-*Unh!* Paving stones! Not meant to be a bike path, I'd say. Sorry, scrotum, multiple insults in a single night. Make it up to you someday. Whoops, path ending. This is more like it: a shady boulevard under overhanging trees. Turn right . . .

Ack! No! Japanese drive on the *left* side of the road, like the English. Fast U-turn; good thing no other cyclists were flying by. Fool of a North American, imperiling local lives. Here's a man walking his dog. Oops, now he wants to chase me; the dog, I mean. Master yells and calls him back.

What a pretty road. Sophora Walk! I can imagine Wang Wei strolling here with his short cane under one arm, composing poems in his head as he walks with his younger brother to the Peach Tree Spring. Okay, I can do that. Here's for you and all reformers, Dr. Sakamoto:

On the Need for New Ideas
Frail plant-tendrils wrap
Black bitter iron. Yet years
Dissolve iron first

And do my ears deceive me? Cicadas? Try this, then:

On Those Who Resist New Ideas
Even in the dark
Cicadas whine and grumble
Enjoying the gloom

The path takes a leftward sweep, merges with another path that comes in from the right, and skirts a little mere whose southern half is filled with rushes. It's a miniature place, a bonsai of a lake, hardly more than a pond; but the views across it are exquisite. I pass an old lady on a bike even older than mine. We nod at each other; she grave and sober, me grinning like a fool. From the time I boarded my JAL flight, the Japanese I've encountered have made me feel at home in a way I rarely do even in Vancouver, and never while traveling before this. There is welcome built into this country's bones; I could kiss the earth beneath my bike tires. *Ting-ting-ti-ti-ting!* Ah, foreign cyclist!

Back at *Sakura-kan*, I park my bike, pat its seat as if it were my small son's head, and reenter the guesthouse. I tender my *cardu*, get my key, find my room, reach into my dorm fridge, and pull out one of the Kirins I'd sequestered there. My oath: perfect. Glacial. Every cell in my mouth yells applause. I carry my beer onto a little balcony and listen to the warm still night, a misty rain now falling, rattling insistently on dark-green leaves: the warm sweet rain of Japan.

THE JOLLY TASKMASTER

As the next day wears on, a vile joke heard long ago comes to mind. An overweight man, having failed at every other diet, visits a Sure-Fire Weight Loss Clinic. The first day he's shown into a locked room where he's greeted by a beautiful woman in a Versaci gown. "If you catch me," she says, "I'm yours." She leads him on a high-speed chase that causes him to shed ten pounds. The next day the woman's wearing a bikini, and in the chase, he loses twenty pounds. The third day he's shown into a smaller room where a 500-pound sumo wrestler, stark naked, is smiling wickedly. "If I catch you," the wrestler says, "you are *mine*."

Dr. Masumi Asakawa is compact, muscular, and jolly: He doesn't look at all like a sadistic sumo wrestler. But the attitude? That's the same. I'm here to interview as many people as I can, to learn as much as I can, right? So let's go! And Masumi sweats me like the track coach from hell. Done this interview, *Bill-san?* Excellent! On to the next one! No need for a coffee break, I'll bring you some tea and you can sip it as you write. Every half-hour, too—not like those leisurely hour-long extravaganzas that Dr. Sakamoto organized yesterday. Quick, now! Down this corridor, through these doors, up this elevator, across a bridge, under a walkway, a shortcut through this lab . . . .

By mid-morning I've sprinted so far that I'm practically tripping on my tongue. Somehow my brain rises to the occasion and functions automatically; when I review my papers on the homeward plane I find I've got eighty pages of closely spaced notes; but darned if I remember writing half of them. I've just run headfirst into the famous Japanese work ethic, and I'm about as prepared for it as a high quad is for a marathon. That's not my last surprise, either. At the end of this exhausting trip I will encounter the Japanese *play* ethic, which will prove even more hazardous to my health. But I'm getting ahead of myself.

I've been dealing with Masumi via e-mail for almost a month; he's a buddy of Yasu Tanaka, the inorganic chemist I met back home. Yasu called Masumi, Masumi called in the heavy artillery in the person of Dr. Sakamoto, and Dr. Sakamoto lined up interviews for me in Tokyo and Tsukuba. Although Masumi has proven to be the key, he now exacts payment for his services by subjecting me to his patented weight-loss program.

The first thing that hits me is his breakneck walking speed. I like to think of myself as fit. In Vancouver I'm the fastest guy on the sidewalk; no one but a cycle courier passes me. But Masumi's personal propulsion system is supersonic. I've jogged at slower speeds. I also pride myself on my spatial sense, and often boast I can find my way around any locale; but now I'm hopelessly lost. It's due to the speed, you see; I don't have time to note the landmarks. I feel like one of those kids you see towed bodily behind a mother running errands in a mall. Masumi's a jolly soul, impossible for anyone to dislike, but I have never had a tougher taskmaster. I think as we tear along, *I'll thank him for this when I'm back in North America.* At the moment, however, it's all I can do to keep breathing.

Masumi blasts through a set of double doors and delivers me into a makeshift office recently carved out of a converted warehouse. An immensely tall and gangly man shyly offers his hand and introduces himself as Dr. Jong Hwa Jung, a scientist visiting from Seoul University in South Korea.

Elsewhere I've remarked on the tendency of nanoscience toward the international; in this it's like all science, only in an exaggerated way. I can imagine research being done under a shroud of secrecy in a classified field like bioweaponry, but stand-alone nanoscience would be a contradiction in terms. The discipline is so new, and is discovering so much so quickly, that any lab isolated from the international mainstream would fall behind in weeks.

Yet at the same time I've observed that strong national traits everywhere flavor, or even determine, the broad thrust of regional nanotech. Let me cite just one example. One could not imagine anything remotely similar to the San José conference occurring in Japan, anywhere, ever. There isn't a Japanese I met, whether in business or academia, who would not gladly undertake *seppuku* rather than behave in public like some of the weirder NanoFornians I met. Maybe it's because I've lived in reticent Canada, but I must say I side with the Japanese.

Concerning those national nano-characteristics: Japan's strengths are dedication, foresight, determination, duty, and resolve. Those of the United States include absolute confidence and unchecked imagination. Both attribute-sets have their virtues. Japan can stay a course through years and decades; the USA can spin off wild new ideas and then marshal the *chutzpah* to see them through. Yet national approaches can also have national drawbacks. To date, as Dr. Sakamoto admitted to me, the most prestigious institute of applied science in Japan has compiled a "miserable" track record in converting basic knowledge into revenue. Conversely, the unfettered U.S. approach can scare off investors with its strain of kooky, quasi-religious hypotheses and outright charlatanry. Perhaps there's an ideal middle course, what the mediaeval logicians called a *tertium quid* or "third thing," but if any nation's come up with it I have yet to see it. Japan and the United States define the extremes, cautious and conservative on the one side and hell-bent on the other, and all other national nanotech programs fall somewhere along a gradient between these poles. Switzerland is like Japan; Canada and Australia are halfway between the U.S. and Switzerland; the U.K. is closer to the American approach. It's a modern case of social Darwinism. We shall shortly see which nation, in nanotech terms, proves fittest to survive.

Another clash between national push and international pull in nanotech comes packaged in the tall, timid gentleman who now shakes my hand. Japan and Korea have a love-hate relationship as deep as that of France and England, or Canada and the United States. Centuries of history, of wars fought and alliances made and broken, both unite and divide them. Japan coined the term "Hermit Kingdom" for Korea, a term that's still strikingly relevant to the insular North. Korea has on more than one occasion attempted to conquer Japan. In fact, the term *kamikaze,* applied to the airborne suicide bombers of WWII, refers to the "Divine Wind" that dispersed a Korean invasion fleet that was bound for Japan in mediaeval times. As often happens when two vigorous but distinct cultures lie side by side, each has inflicted cruelties on the other. Japan and Korea are officially at peace and are even tentative allies under the region's U.S. military hegemony, but some strains of mutual mistrust linger on. The Koreans remember the wartime enormities of sixty years ago, when they were a fiefdom of the Empire of the Sun. In late 2002 North Korea made the astonishing admission that a few years ago it kidnapped dozens of young Japanese, brainwashed them, and set them to work in its spy program. The Japanese are exasperated with what they see as unrelenting demands to offer groveling apologies for actions for which few alive today bear direct responsibility. Both Japan and South Korea are thriving democracies with modern economies, says Tokyo: Surely we can look forward instead of back?

Both sides are making progress, but there are miles to go. When their team was defeated in the 2002 World Cup, many Japanese rooted for South Korea's Reds. At the same time, they doubted privately whether most Koreans would prove as generous had the Reds been eliminated instead of the Japanese.

But nanoscience is among the most international activities on earth; and so in Tsukuba, I found Japanese and Korean scientists working side by side to derive new data on the lab bench. Perhaps the surest way for these two nations to create close ties will come from many such individual instances of collegial cooperation among nanoscientists. The laws of physics are everywhere the same—even in Korea and Japan.

Dr. Jung is a researcher in Masumi Asakawa's nanoarchitectonics unit. His business card bears this cryptic description: *CREST (Toshimi Shimizu Team): Functional High-Axial-Ratio Nanostructure Assembly for Nano-Space Engineering.* Right, I say, as Masumi and I brake to a halt and my heart rate starts to slow from 180. What is it that you do?

"Before I answer that, I must give you some background," Dr. Jung says half-apologetically. Like everyone I've interviewed here, he deals gently with my ignorance, as if I were the boss's idiot child. "When you consider inorganic molecules, titanium dioxide and the like, you see they arrange themselves in fairly simple patterns. By contrast, organic molecules have much more complex shapes. When you look at a CN, a carbon nanotube, you see that its shape is simple. What does that suggest? That it is inorganic, yes? Despite being entirely carbon. It is more like poor dead diamond than a substance precious to, or produced by, life.

"My area of investigation is *non*-carbon nanotubes. What? Oh, yes. Nanotubes do not have to be made of carbon—quite the contrary. I look at silica nanotubes. These are self-assembling structures that have many uses, including the catalysis of certain organic reactions.

"I have been able to get a non-carbon nanotube, twenty nanometers in diameter, to acquire a helical pattern that I predetermined. It self-assembles to virtually unlimited lengths. In certain instances I can get it to achieve structures that look almost like synthetic bone."

The actual mechanism of self-assembly, Dr. Jung goes on to say, is almost exactly like the lost-wax process by which an artist's foundry casts bronze positives from plaster originals. "We can deposit certain metals, silver or palladium, inside a silica nanotube, atom by atom. We can space these atoms as regularly as soldiers on a parade ground. Commercial uses?

Oh yes, we think there are many. We can construct made-to-measure molecules that function like the active sites of natural enzymes. Synthetic catalysts, you understand. These nanoscale regions favor the formation of desirable end-products. Different isomers of quartz can be used in this way to produce specified alcohols with 100 percent efficiency.

"Another area of interest is using carbon nanotubes as storage canisters for diatomic hydrogen, H_2. If we use cryogenic storage—that is, storage at supercold temperatures—we can store up to six percent hydrogen by weight. That falls to one percent at room temperature. Or we can use other types of nanotube, multilayered silica for example, and get room-temperature hydrogen storage up to almost four percent at room temperature. This effect will prove to be very important whenever the world turns to hydrogen instead of petroleum as its universal energy currency. It is all very well to speak of cars and trains being operated by fuel cells. But they will need the equivalent of gas tanks, will they not? Someplace to hold the hydrogen fuel they use. We think we are on to a means of meeting this future commercial need."

I'm still digesting all this when Chief Taskmaster Masumi again takes charge of me and whisks me down another maze of hallways to meet Dr. Takeshi Sasaki.

Within AIST, Dr. Sasaki's unit looks at something called high-interface-area nanostructures (HIAN). That means he and his colleagues make materials that act as complex devices: substances, even individual molecules, that are true machines. Their new materials can do this because at the nanoscale, they have been engineered with specific functions in mind.

Dr. Sasaki has any number of examples. Here is a material that has been grown to be honeycombed with 3-nm passageways—what Dr. Sasaki calls "nanopores." Gas molecules of a specific type, and no other molecular variety, infiltrate these nanopores and react with receptor molecules that have been built into the sides of the nanopores a few angstroms from the material's surface. The result is a gas detector that can be made as an ultra-thin film and deposited on any substrate. No need for big, clunky, delicate, expensive devices like gas chromatographs. Not now. Nanotech has come up with a paintable sensor, as easy to use as litmus paper.

Looking at Dr. Sasaki's HRTEM photonanograph, a cross-section of the new material, I utter a surprised laugh of recognition. The thing looks exactly like Japan's ubiquitous *pachinko* game. In this national pastime, addicted gamblers fill vast parlors and try to guess the pathways of little spheres tumbling down through intricate passageways. National nanotech, indeed!

"Our approach is to fuse inorganic materials with metals," Dr. Sasaki tells me. We're strolling along the hallways outside his labs proper, looking at poster displays of the technology he develops. "Nanotechniques allow us to build in additional functionalities beyond the microscale properties of standard composites." Say what? "I mean we can tailor-make a material." His HIAN unit can even get photoconfinement effects. I'm about to ask for clarification when Masumi shakes a finger at me. "We are coming to that!" he says. "Your next interview will deal with photoconfinement and plasmons." Chastened and instructed by my parole officer, I move on. What, I ask Dr. Sasaki, is his preferred means of making these new nano-structured materials?

"We use sputtering to a large degree," he says. "This involves heating a source material in a high vacuum. Atom-sized bits of material are boiled off and then self-assemble in regular layers on a substrate. If we sputter two different source materials at the same time, a homogenous mix appears and we end up with a self-assembled nanocomposite. We can also deposit alternate layers of inorganic films and nanoparticles, while controlling the nanoparticle size and the film thickness. This permits us to produce nanocomposites whose mechanical, electrical, and thermal properties are not usually associated with composite materials. Via nanoengineering, these traditional effects can be substantially adjusted."

Adjusted to what? Dr Sasaki grins. He has some surprises for me. "It depends in part upon the nanoparticle, and in part upon the matrix," he says. "See this chart." He points at a grid on the wall:

Nanoparticle	Matrix	Function
Silicon, Carbon, Germanium	Silicon dioxide	Photoluminescence
Silicon	Magnesium oxide	Photoluminescence
Cobalt oxide	Silicon dioxide	Optical gas sensor
Silver iodide	Silicon dioxide	Photochromism
Platinum	Titanium dioxide	Photoelectrodes & Photocatalysis

"What do you mean by photochromism?" I ask, scribbling furiously.

"Self-darkening sunglasses would be one example," Masumi answers. He's leaning against the wall beside the hall poster. "Reversible chemical change due to the presence or absence of light."

"And photocatalysis?"

"One-way chemical reactions made possible by light. Irreversible reactions."

"So your nanomaterials—"

"They do things *innately*," Dr. Sasaki says, brandishing a pencil in the air. "These new substances of ours perform various functions in and of themselves. We expect this will revolutionize many kinds of industrial processes."

DR. JUNJI TOMINAGA is spare, slim, elegant, and precise. He's affable enough, but he seems, well, *controlled*. Even within the best and the brightest of AIST, he's a heavy hitter. As well as being director of the AIST Laboratory for Advanced Optical Technology, Dr. Tominaga holds appointments as full professor at Tokyo Denki University and visiting professor at Cranfield University in the U.K. In addition, he maintains close contact with seven Japanese industrial titans including JVC, TDK, and Toshiba. Some academics may inhabit brain-coffins, but not this man.

Dr. Tominaga is an expert in optical near-field technology. This discipline uses an odd variety of photons—those fundamental bits of light that act as either wave or particle, depending on the context. Dr. T's strange photon variants are called plasmons.

Normally photons have only one state: They move ahead at the fastest velocity known to physics—namely, the speed of light. Photons have a tiny mass, and only for relativistic reasons; they weigh something simply because they go so fast.

Not only do photons illuminate the universe, they hold it together. The electromagnetic force, which knits protons to electrons and makes possible all atoms, is mediated by photons. Photons continuously shuttle back and forth between two charged particles, making them attract (if they are of different signs) or repel (if they have similar signs).

Under certain conditions, a photon can get trapped in a material surface, usually a metallic one, and remain bound there as a standing wave. Unable either to pull free and resume its high-speed travel, or else to burrow more than 100 nm or so into its host material, this bizarre, atypical photon form is called a plasmon. It was utterly unexpected, having been unforeseen by classical theory. Thus a plasmon is another case for basing nanotech on what *does* exist, not what *should*. My mental image of a plasmon is a drop of water dancing on a hot skittle: It defies predictive logic, but there it is.

Dr. Tominaga believes plasmons, for all their oddity, could prove an excellent avenue toward superdense optical storage of computer information. Plasmons, he thinks, could be created and stored, then read or erased, as both RAM and ROM.

"It appears," he tells me, "that we can modify existing DVD technology by replacing the customary red laser, which has a long photon wavelength, with a shorter-wavelength blue laser. Ancillary modifications would be necessary to accompany this change. For example, red lasers can be focused using plastic lenses. A blue laser would explode a plastic lens—the energy coupling between lens and light would be excessive.

"If a blue laser were to create plasmons on the surface of an optical disk, and if we could get it to write, read, and erase these plasmon-based data, then our theoretical limit for data storage would approach fifty or one hundred gigabytes per disk. This is ten to twenty times the current limit for magnetic storage." Um, I say. Didn't AT&T come out with a one-terabyte disk ten years ago?

"Not really. They did achieve high data densities over a small area, but the whole disk area was never filled, so we cannot strictly speaking call it a 1-Tb disk. They indulged in—shall we say—unjustified extrapolation."

Dr. Tominaga thinks the key to plasmon memory is found in one of his team's recent innovations, the super-resolution near-field structure, or super-RENS. "I have taken a certain amount of ribbing for this term," he says with a tight smile. "People, Westerners, think I am trying to say 'lens.' I do not care, as long as the term sticks in people's minds." Another reason for it to stick: The super-RENS was granted U.S. patent 6,226,258 on May 1, 2001.

Tominaga *et al.* install a super-RENS on the surface of a standard polycarbonate video disk by using dry sputtering methods developed by Dr. Sasaki and others. It's a textbook example of the kind of inter-lab, intra-agency collaboration that Japan hopes to maximize with its reorganization of AIST. Using these methods, Dr. Tominaga's team lays down silver-oxide particles on the DVD in regular stripes 100 nm thick. Striated in this way, the AgOx particles do not clump together. The dry deposition process holds them to a uniform 20–30 nm diameter. This size makes each particle ideal for hosting one data-storage plasmon. Using plasmons, the silver nanowires can store RAM data more effectively than today's best magnetic methods.

"We used standard DVD recording technology to change the nanostructure density of the silver-oxide particles," Dr. Tominaga says. "We then were able to demonstrate super-resolution characteristics arising from the plasmons in the silver film." A single 100-GB disk, which Tominaga *et al.* believe they are on the brink of perfecting, could store a hundred 90-minute movies—a video library in your pocket.

ANABOLIZING ALCOHOL DEHYDROGENASE

All laboratories smell alike. Blindfold me, drop me down in an org-chem lab in Tsibili or Tsukuba, and until I heard someone speak I wouldn't know where I was. Maybe not even then, given nanotech's international interconnectivity and prevalence of English as a common denominator. So when Dr. Masumi Asakawa finishes his patented weight-loss program and ushers me into his own laboratory for my final interviews in AIST, with my first whiff of air I'm instantly at home. This could be Toronto, Texas, San José, or McMaster—anyplace I've visited on the long, meandering path my research has led me. I close my eyes, inhale the aldehydes, open my eyes again—and am totally unprepared for what I find. Except for Masumi and myself, this lab is populated entirely by women.

There are a few high-profile female researchers in Japan. Dr. Midori Sowai at the Tokyo Science University, for example, was spoken of in reverent terms by Dr. Jong Hwa Jung for her pioneering work in quartz catalysis. But until this instant, the only women I have met within AIST were at reception desks or serving tea. Suddenly this has changed. Until this moment, I have not realized how intensely I missed the feminine in Japanese science. Masumi beckons me into his stronghold with a big smile.

"These are my team!" he announces proudly. "I am encouraging them to show their work to you. First, to practice their English, which is necessary in the world arena. Second, to get used to the experience of presenting material before colleagues from abroad. You will talk first with me, then Dr. Hiroko Yamanishi, and finally with Miss Megumi Akiyama. Megumi has her M.Sc. and is working on her doctorate."

Fine, I say, falling onto a lab stool, but first I need a coffee. "Done!" says Masumi. He swivels his chair around, reaches for some implements, and gets the stuff himself. It's instant plonk, but given my state of fatigue it tastes like nectar. Tell me what you do, I ask.

"We produce molecular devices based on organic materials," Masumi says, grinning. "My postdoctoral work, which I did in England, was on chemicals called porphyrins. They and their derivatives are what I still working on today." Porphyrins? Like what Neil Branda studies back in Canada? "Yes. This is cognate with Neil's work.

"A chemical called catenane, for instance. It's a very interesting molecule, configured as two rings that are closed and interlocked. You could use this in nanoscale machinery as a bearing. Or as a molecular motor, provided you could get it to turn the way you want. Right now it spins any which way, driven by the Brownian motion of adjacent molecules when it's in solution. But we think we know how to drive it in one preferred direction. Any reducing chemical will donate protons and create a reaction that spins catenane at a steady 15 hertz, or 900 rpm. We have filed for a patent on this exothermic reaction.

"There are also other molecules called rotaxanes. But I'll let Megumi tell you about those—they're her specialty. She's about to publish a paper on the topic.

"As well as the rotational motion of rotaxanes and catenanes, we're looking at other molecules with reciprocating straight-line motion. That's piston-like motion, the kind that actuators have. How do we handle them? Well, we splice fullerenes to these molecules. The fullerenes act like doorknobs, making the attached molecules much easier to locate, grasp, pick up, and move around.

"I don't want to make this sound too easy, Bill-san. There are always problems. One particular difficulty in studying molecular motors is how to obtain directly observed data. We want to study a molecular motor, a rotating molecule, so what are we forced to do? Immobilize it! Freeze it, so we can inspect it in a scanning probe microscope. But if it's immobilized it can't spin, right?

"Perhaps in the future, the new STMs [scanning tunneling microscopes] can be tweaked to give us very fast data-capture times. So in theory, we could take a series of successive images at part-second intervals and string them together to create a video of molecular operation.

"That's one of the new methodologies we're working on. We're also devising a technique that lets us anchor only one part of the molecule, leaving the other part free to spin." Hah! I say: stator and rotor, just like a car's disk brake. "Oh, we have brakes, too!" Masumi says, bouncing on his stool with glee. "Megumi! Tell Bill-san about the brakes."

Megumi Akiyama swivels around from a workstation where she's been crunching data and playing Aerosmith. She's silent and unsmiling, but not hostile: just as calm and focused as a blue-laser beam. Physically she's no bigger than my eleven-year-old boy, but an aura of confidence and power surrounds her so completely that it's only later that I realize how tiny she is. Women like this have governed empires.

Megumi's speech is initially hesitant, but Masumi's decision to give her practice is a good one. Her English quickly grows more fluent with use. Masumi sits nearby, fielding answers if the lady is stuck for a phrase and so sparing her embarrassment. But in no way is he hogging interview time. Unlike many lab directors and politicians who take credit for anything good that occurs during their tenure, Masumi accords his younger colleagues total credit for the work they have done, and which they now describe. The man is the best boss I have ever seen.

"There are many types of rotaxane," Megumi says. "We use the simplest type to minimize problems. Rotaxane and catenane have high motility and free rotation. As well as using these molecules for motors, we think they could be used also for switches." She's a little hesitant on the liquid sounds, but distinguishes her Rs and Ls quite nicely.

"Rotaxane and catenane can self-assemble," she tells me. I give a small grunt of surprise because I hadn't known that. "Synthesis proceeds through an intermediate molecule called pseudorotaxane. If these molecules are modified so that some of their carbon-carbon bonds are covalent, they lock up: There is no rotation." These are the brakes that Masumi was talking about, then? I catch his eye; he nods. Megumi continues: "Rotation speed of the freely spinning molecule varies according to temperature, over a range of one hundred to one thousand hertz." I lift my eyebrows: 1,000 Hz is 60,000 rpm—ten times as fast as a Detroit motor's redline. How, I ask, can you make such a molecule rotate in only one direction?

"The radial vector, you mean? We are looking at ratchet effects. This is what powers the ATPase motor in a bacterial flagellum. The motive molecule expands and contracts repeatedly, forcing the rotor around." Biomimicry again, then? "Yes, certainly. Of course." Empress Megumi inclines her head in a single gracious bow, then glides back to her Aerosmith. The interview is over. I'm left feeling I should tug my forelock and ask permission to mow her lawn. But there's no time for this, as Dr. Hiroko Yamanishi now takes Ms. Akiyama's place. Masumi beams indulgently as his second star youngster takes the stage. The second-rate, it's said, surround themselves with the fifth-rate: hence most governments. But the first-rate choose other first-rates, always.

Dr. Yamanishi at once launches into a detailed explanation of organic-chemical synthesis. This leaves me as breathless as her boss's hall-sprints. With head-spinning speed she covers new nanotechniques that create rotaxane brakes; recycle potent and expensive chemicals such as crown ether (instead of using them once and then having to dispose of them); and make molecular machinery self-assemble so that Drexlerian nanobots are as needless as a referee at a lovemaking.

The recycling methods, Masumi interjects gently, are something new.

"This is green chemistry," he says. "The start of it, anyway. For years chemistry has taken abuse from eco-activists. They say that chemistry creates unnatural compounds never seen in nature and dangerously foreign to our immune systems. Now chemistry is about to show that it can do more with less—less energy, less input material, less waste. Go on, Hiroko."

And Hiroko does. The details threaten to overwhelm me: electron concentrators, proton donors, UV spectral analysis, direct optical imaging, gel polymerization chromatography, collision-induced dissociation . . . After a half-hour my head feels as if it's surrounded by closely orbiting bluebirds, but one phrase my interviewee has used sticks in my mind. I ask her about it: *polymeric synthesis.*

"Oh, yes," she says, as if it's too obvious to mention. "I'm sorry, did you not understand that? We are developing a completely new method of synthesizing polymers. Dyes, drugs, detergents—in a short time we will know how to make all these things faster, with less waste of energy and raw materials. That is the whole purpose of my work here. This is our next big application in nanotechnology."

Masumi Asakawa beams approval from an adjacent chair.

THERE IS, Dr. Tsunenori Sakamoto told me earlier, an outer Japanese. And then there is an inner Japanese. My visit to this remarkable place called Science City is about to end with the latter. You can't beat local contacts, that's the moral here. Without Yasu, Masumi, Dr. Sakamoto, and all their introductions, I'd still be sending desperate, formal e-mails from Vancouver, pleading for a brief word. Now the doors have been thrown wide. I've already seen enough hints about the "inner Japanese" to reject that earlier misconception about this nation's so-called standoffishness. It's a total slur. Everywhere I've been received with frankness and professionalism, certainly; but with cordiality as well. I have never traveled, even within North America, and felt so at home. Now, as an even greater compliment, I am shown the inner Japanese.

Friday night I finally finish at Masumi's laboratory and walk the half-kilometer to the guesthouse. I pat the seat of #25 Lending Cycle (the true Westerner attends to his mount before himself) and go up to room SB36. It's the way it always is: tiny, serene, lit softly by gray dusk light. I hang up my suit jacket, see the room slippers left for me in the closet, and smile. *Sakura-kan* thinks of every detail, and one of these thoughtful details is a comfortable, disposable pair of slippers for the tired feet of the honored guest. At least they look comfortable; I wouldn't know—I can't get them on. My oversized U.S.-Canadian splayfeet get stuck so badly that they hang out the rear end of the slippers from arch to heel. I need ten of the things, one for each toe.

I lose the damned rumpled suit, anyway. The shower feels fabulous; my brain starts to work again, recovering from its overload. Towel down. Now, what to wear? I settle, reluctantly, on half of my earlier getup: blue suit pants and Florsheims. I top it off with a blue checked dress shirt that I leave open-necked, and hope I won't appear as out of place as I've felt all day. Luckily, I've been caught in the rain so much that my pants have lost all shape. They aren't casual, but they may as well be wool jeans. *Every girl crazy 'bout a sharp-dressed man.*

Masumi meets me downstairs. Where are we going? I ask. Lunch was at a Mexican restaurant: great grub, but a cultural disconnect. The walls festooned with grainy black-and-white photos of banditos, the waitresses like geishas dressed for Halloween. *Ay caramba, señor Bill-san!* "Don't worry," Masumi tells me. "Tonight we'll show you the real Japan. There's a noodle house over in Tsukuba's entertainment quarter where I'm good friends with the owner."

We pick up Masumi's wife, a lively, pretty lady who works as a wedding planner. The noodle house is tiny and low-roofed, with odd-shaped rooms opening out in unpredictable directions. People start arriving. Everyone in Masumi's lab, together with their boyfriends; a Korean graduate student; a high-school student who's staying with the Asakawas. And others, and *others*. By eight o'clock there are twenty people jammed around a midsized table. We have a saying here, Masumi tells me: *Let's go from beer.* The local brew is excellent, crisp and hoppy, the temperature of melting ice. Then the food starts arriving. No one orders it; the host just does a head count and then starts delivering what she thinks you'll like. Try this, Masumi says, it's soft bone. There are raw snails in the shell, alive for all I know, that I pull out with chopsticks. At one point I'm eating something soft, gray, and delicious. Do you know what that is? Masumi asks, grinning. No, I say, and don't you dare tell me.

Masumi's friend is complaining about something. What is it? I ask, and he pulls up his sleeve to show me. His cat attacked him, Masumi says. American shorthair, adds the man. I tell him he's a lucky man. An American cat could have pulled a gun.

And then the lighting of the lamps. The *sake* arrives.

Do you know sake? Megumi asks. I nod. (Do they think I'm a barbarian?) I have it at home; often I heat it. What brand of sake? asks one of the lab-husbands. When I tell them, the table explodes in a collective groan of anguish. Apparently my brand is considered two cuts below liquid shoe polish. No, no, Masumi says, dead serious this time. You must drink *sake*. Try this. He pours me a tiny amount in a pretty ceramic cup hardly bigger than a thimble—three sips, the Japanese say. I take a little on my tongue.

Well, sing to me, ye angels. This stuff is like the distillate of springtime, like blossoms in an orchard on a sunny day. It's incredibly light and delicate, there and not-there like cotton candy, its tastes and fragrances hinted rather than expressed. It makes any other drink I've ever had seem foul. Omigod, I say when I can talk again. Forgive me, I didn't know.

AT NINE ON Saturday morning, I'm doing penance. It is, I suppose, a necessary correction for being treated for the last four days like a head of state. I'm schlepping my luggage through a dense, steady rain to the main gate, where I can catch the *highway-busu* back to Tokyo. I won't say I'm hung over: The sake was too well-mannered for that. Let's say my liver has temporarily exhausted its ability to anabolize alcohol dehydrogenase. But, oh, the memories. The toasts, the jokes, the pledges, the abdominal muscles that eight hours later hurt from laughing. And I'd thought I was too old to make new friends.

THE ROAD TO SELF-ASSEMBLY

Tokyo University is a fascinating place. It's three kilometers north of my hotel, right next to *Ueno-koen,* Tokyo's biggest concentration of parks and art galleries. By a miracle the area escaped Allied saturation bombing in 1945. While its older buildings have begun to crumble under the newer, slower scourge of acid rain, the great trees of the university remain, shading wide boulevards and making the place seem like the Harvard of Asia. I'm here to interview the world's leaders in the nanotechnology of self-assembly.

Dr. Masaru Aoyagi and his thesis adviser, Dr. Makoto Fujita, are singularly adept at getting complex geometric structures to bolt themselves together at the nanoscale. Dr. Aoyagi's announced intention is to one day get an iMac computer to assemble itself. His current study area, he tells me, is "guest-induced assembly of coordination nanotubes." I ask him what this means.

"There exist linked sets of molecules called hosts and guests," Dr. Aoyagi explains. "They are fascinating chemical systems. A guest molecule is a small independent entity that sits wholly or partly inside a larger molecule, the host. In certain cases one can get a guest to function as a mold. Around this mold a host molecule may self-assemble, taking the host's shape on its inner surface."

Using such methods, Dr. Aoyagi has learned to assemble large, hollow, tubelike hosts in fast and effective ways. Previous techniques often started with ring shapes, each ring a cross-section of a finished tube. While these could and did self-assemble into a pipe, the pipe's length was uncontrollable. Sometimes it would go on adding sections on either end, headed out to infinity until it exhausted its raw materials. Dr. Aoyagi's new method starts with three C-shaped plates of set length, each plate making up one-third of the finished tube. When these three modules unite, lo! A finished host molecule of predetermined length is the result.

Dr. Aoyagi's nanoscale control of these events is absolute. The C-shaped plate, for example, is a synthetic molecule of his own design and fabrication. I suggest "aoyagite" as a formal name; he blinks at me as if I've spoken a foreign language, or at least been grossly impolite. Evidently it's never occurred to him to immortalize himself in this self-aggrandizing way. How, I ask, can he get matter to behave like a Meccano set—and one that bolts itself together to boot?

"Metals are the key," Dr. Aoyagi tells me. "Metal ions have precisely defined bond angles. Palladium, for example, has a strict bond angle of ninety degrees, so that it functions as a perfect corner brace. That in turn gives us a hollow host molecule of square cross-section. The individual flat plates that make up the host self-assemble about a guest molecule called biphenyl carboxylate. It is the formwork, if you will."

The self-assembly is totally reversible, Dr. Aoyagi says. "When we extract the guest molecule in solution with chloroform, the four flat plates of the host break apart. When we remove the chloroform, the host plates automatically reassemble back into the hollow-square nanotube, with the guest molecule inside. We can extend this complex self-assembly to create a very involved configuration, a honeycomb shape, that consists of multiple nanotubes linked side by side in a grid."

Possible commercial uses? "No definitive answer yet, I'm afraid. We're just looking at the fundamentals of how nature works in geometrical systems—though this approach may well prove useful in optimizing common processes for industrial chemistry. It could make them faster, simpler, and more likely to go to completion. So, yes, this is basic work—nanoscience, not nanotechnology. But it's just this type of basic work that created great chemical conglomerates like DuPont."

If Dr. Aoyagi is a young prince of nanomolecular self-assembly, his thesis professor Dr. Makoto Fujita is an undisputed king. Along with Dr. Jean-Pierre Sauvage at l'Université Louis-Pasteur in Strasbourg, France, Nobel laureate Dr. Donald Cram of the University of California, and Dr. Julius Rebek of La Jolla's Skaggs Institute for Chemical Biology (part of the Scripps Research Institute), Dr. Fujita has made a name for himself in what he calls "directed self-assembly." It was Dr. Fujita's brainstorm ten years ago that led to the use of metal ions as structural corner braces. *Science* magazine has compared this discovery to the ingenious invention of flat cylindrical connectors that made possible the children's construction set called Tinkertoy.

At the moment, the Fujita team is exploring another of their inventions. It's a nanoscale octahedron, a molecule shaped like two four-sided pyramids glued together base-to-base. With an inside diameter of only 3 nm, the Fujita octahedron makes a nice snug container that exactly fits a C_{60} buckyball.

"If you try to use conventional synthetic chemistry to make such structures, you won't succeed," Dr. Fujita tells me. "But if you use our approach to nanochemistry, directed self-assembly, it is easy—almost absurdly so."

It's nine o'clock on a rainy Saturday evening. We're sitting in Dr. Fujita's big, dark university office, flanking a computer-driven projector that shows a series of complex molecules fitting together. Given the time of night, it's not surprising we're the last ones left in the building.

Another term for what Fujita et al. are doing is "paneling," he explains. "We make these large, flat organic molecules that are thin and rigid. They're shaped like slabs of plywood, and we treat them as such. We can put them together, or rather get them to put themselves together, in many different ways. Sometimes the hollow inside of the structures we create seems to work like the active site of a natural enzyme." How so? "It catalyzes certain chemical reactions." But won't these reactions get out of hand? "Oh, no. We simply close off the reactive parts of our new molecules with inert caps." How large are these overall structures? "We have reached outer diameters of five nanometers and molecular weights of 20,000. Those are the characteristics of a fair-sized organic molecule like insulin."

As advanced as this is, Dr. Fujita explains, it's only humanity's tentative first step toward a workable chemical biomimicry. He changes the slide to show a long-chain molecule coiling itself into a nanometer-sized cylinder. "Here you see a harmless natural life form called a tobacco mosaic virus. It is a perfect example of advanced molecular self-assembly. Its repeated construction unit—its wall module, so to speak—is a small protein. The virus clips together many of these small proteins into a long polymer called a polypeptide, which constitutes the viral shell. The shell occupies a borderline, a gray area, between a self-assembling chemical compound and a living system.

"Our approach to directed self-assembly takes lessons from nature. We must learn from nature, because it has more elegant methods of building things than we have so far come up with. One of our group's goals is to do what a virus does—to easily and quickly direct the self-assembly of complex structures."

Despite Dr. Fujita's modest statements about merely imitating nature, some of his team's structures have few or no natural homologues. They are wholly new under the sun. It is not nature's structures that he wishes to mimic as much as its methods. In my view, this makes his team's activity the most powerful invention since biotechnology or atomic power.

"We can create large rings of DNA," Dr. Fujita says, "by bending around the ends of long DNA strands and joining them. We can catenate these rings, or interlock them, so that they associate together like the links of a chain. We can make and break this chain at will; further, we can do so automatically. Its occurrence is spontaneous. We merely mix the right chemicals, create the right environments, and watch it happen."

Create the right environments. The phrase arrests me. Where have I heard it before? Then I remember: Tom Theis of IBM, speaking at San José. It's the final head-whack of my research: the great connection, the completion of a quest involving a hundred scientists and nearly as many disciplines. Dr. Fujita has realized from chemistry what Tom Theis, whom he's never heard of, has independently seen via computer science. The two areas of study are so different that their practitioners rarely read the same journals or attend the same conferences. But while they have come through different doors, they have arrived together at the same place. Here is the profound truth they have uncovered. Call it the Fujita-Theis Rule: *A living system packs minimal information, one of whose central functions is to extract from the environment the additional data it needs but does not itself possess.*

Self-assembly occurs because the environment and the living system it contains act in concert. They are allies, working toward the goal of ever more complex systems. No, that's too mild a statement. Life and the environment are a single thing: an interlocking system. Nietzsche was right—the universe moves toward self-overcoming of its own free will.

Complexity, pattern, the creation of new things, all arise as surely as night follows day. When people ask me why I write books, I tell them about times like this—the question answered, the circle completed, a flash of silent lightning in the brain. And always another iteration, with more penetrating questions. What a rush.

It strikes me, too, that this is the final nail in the Drexlerian coffin: the last thing that relegates Drex and the Dreamers to the waste bin of rejected religions. Dr. Fujita, with colleagues in France and elsewhere, has produced some fascinating molecular machinery; but that's not the point. The crux is that Dr. Fujita *et al.* have got this complex stuff to *assemble itself.* Once again, this time from a totally different perspective, I see why all those Drexlerian plans, the cross-sections and straight-faced engineering diagrams, are so ridiculous. We don't need a fiendishly complex toolkit to manipulate the nanocosm; we need only learn to make the nanocosm manipulate itself. Not *make,* rather, but *let.* If we give the nanocosm the right conditions, all we have to do is stand out of its way while it effects its miracles *sui generis.* The nanocosm is like Al Capp's Shmoos, critters that simply ache to become whatever you want. A Shmoo will leap into a frying pan if it thinks you'd like some bacon, and as it dies will modify its flavor to your preference. At this point, all science and technology need to learn is how to put humanity's requests into a form the nanocosm understands. Then in microseconds, femtoseconds sometimes, the nanocosm will leap to do our bidding.

Unlike some researchers, Dr. Fujita is quick to discuss possible applications for his team's work. Rotational molecules, for example, could serve as logic gates and memory cells for nanoscale computers. Because of this, he says, the day may come when a child can hold a supercomputer in one hand. At the moment, however, the most intense industrial interest in Dr. Fujita's work is coming from the giant Japanese firm, Fuji Photo Film Co. These days the company is caught in a dilemma that it has very little time to resolve. It has made billions of dollars by producing top-quality celluloid-based film for still and movie cameras. But in today's world of increasingly refined digital image capture, Fuji's current core expertise has become something like making whalebone corsets: When the thing itself becomes outmoded, quality doesn't count. Fuji must decide whether it is an imaging company or a chemical company. If the former, Fuji must abandon chemistry. Then it must play catch-up with firms even bigger than it is, firms such as Kodak and Agfa that have literally decades of lead time in that field. If Fuji Photo Film abandons imaging, it must look for ways to apply its hard-earned chemical expertise. Furthermore, it must do so in bold new ways—again, to avoid head-to-head competition with the older, larger firms who already dominate the chemical sector: DuPont, Monsanto, Morton Thiokol, and the like.

In striking an informal partnership with Makoto Fujita, Fuji has implicitly stated that it's reestablishing itself as a chemical company. We can expect, however, that—given the quality of scientist it is retaining—the chemistry Fuji Photo Film harnesses will be unlike anything the world has seen before.

CONCLUSIONS AND SUGGESTIONS

There's a grand tradition of foreign writers breezing into a place they've never been, spending a few days, and blithely telling the world (including the bemused natives) what's wrong with the place and how to fix it. Since this tradition is hallowed by centuries of use, I will not attempt to amend it. In my defense, this section may be a useful deposition from *l'oeil rinsé*—the fresh eye.

The only thing I would change about Japan is the timidity, uncertainty, and self-doubt I sensed in some of those I interviewed. *This* is the worst smog in Tokyo, a city whose material air is infinitely better than that of Los Angeles. It is a miasma of hopelessness and defeatism that astounds me. Japanese publishers wait to see what instant-sure-fire-gosh-almighty management techniques become fads in the USA, then rush to translate and distribute these materials. There are Japanese readers whose sole intake is me-too books of this sort. Japanese scientists whose research leads the world are ignored by their homeland industry. Japanese industry scorns the very R&D techniques that made it rich.

I hereby highly and holily declare that there is no reason for Japan to adopt or encourage defeatism of this sort. I reject the notion that *if* Japan exerts herself, she *might* maintain third place among the world's great communities—lagging after the United States and the European Union. This is neither the spirit that wrought "the Japanese miracle," nor the spirit to restore it.

A symbol comes to mind: the ultimate national symbol of a country's flag. Japan's wartime banner, the Rising Sun, has been set aside. But what has replaced it? Is it not the *whole* sun, a star fully risen? The nation that arose from its own ashes after WWII, that practically invented the world's knowledge economy, that made the most brilliant use of industrial R&D the world has yet seen, that rose to dominate the auto and electronic sectors, that was the first nation anywhere to see that wealth came from brains rather than resources, that forced the flabby and self-satisfied economy of America to reform or perish, that fought an economic war as courageously as it had fought its shooting wars—this nation has no business relegating itself to third or fourth place. Not now, not ever. Japan's modesty, her unwillingness to intrude into the councils of the mighty, may be admirable manners. But in terms of *realpolitik,* it is suicide. The time for shyness is past.

Shakespeare said it perfectly. "Be not afraid of greatness: some are born great, some achieve greatness, and some have greatness thrust upon them." Japan is of the latter sort. She has already saved herself by her exertions; and if she relocates her core and her genius, she can save the world by her example. Consumers, stockholders, governments, trading blocs the world over can only profit from a strong Japan. As an ally, she will make democracies secure. As a competitor, she will keep us fit and honest. As a friend, she will share her vivid history and exquisite culture to enrich us all.

The key to the Japanese Renaissance is twofold: nanoscience and nanotechnology. These two linked activities are not just another new discipline, good for some Ph.D. theses and a university department or two. Like the transforming ideas of the past, agriculture and automation and calculus and cybernetics, the *Shirotae* of nanotech—the "white mysteries" of creation, the truths beneath Truth—form an entirely new way of looking at nature. And of harnessing nature. Not *harnessing,* rather, but *emulating.* Japan, like the United States, is perfectly placed to show how this can be done. Both national viewpoints are needed. If vigorously pursued, the two approaches can be the negative and positive poles of an enormous, enabling battery: an energy source that at last converts the earth into one place and its fractious nations into one people.

Certain changes are called for in Japan. With humility and trepidation, I offer some suggestions. Japan must have a permanent place on the UN Security Council. An even greener Japan is required, a polity that shows both developed and developing nations how to live lightly on the land. The Japan Self-Defense Force must have tactical nuclear weapons for a vicious sting that dissuades any and all would-be aggressors. All these things are mandatory. But before they can occur, Japan must shake off her hesitancy, don again the armor and honor of her *samurai*, and stride forth with calm confidence and universal goodwill to make a lasting mark in the world. It will be difficult. But it will not be nearly as hard as what this unique and marvelous nation has achieved before.

One clue convinces me that such vital and necessary changes have begun. Dr. Tsunenori Sakamoto showed me an AIST chart that divided nanotechnology into various areas—electronics, smart structures, materials, pharmaceuticals. Each area fell into one of three categories. The first was Japan Dominates in Three Years. The second was Japan Dominates in Ten Years. The third was Japan Dominates in Twenty Years. No area whatsoever was allotted a category for second place.

Hai Nippon!

CHAPTER 10

NANO-PITFALLS

DR. MURPHY AND DR. WOLFRAM

K. ERIC DREXLER portrays his Holy Grail, the molecular assembler or nanobot, as an item of great complexity. Drexlerian nanomachinery is staggeringly complicated—unworkably so, in my opinion. Yet even a nanoassembler's complexity of design and construction pales beside the task of its operational complexity (i.e., actually *running* the thing). To function, a Drexlerian nanobot would have to first store high-level instructional software onboard in large quantity. And to work as the Drexlerians want, the nanobot would also have to distinguish among many possible conditions, materials, and configurations, and then act instantly and appropriately in every case. It would have to sense distances to sub-angstrom accuracy and act in shavings of a picosecond. Given the absurdly tight dimensional constraints, there would literally be no room for error.

A nanobot that fit into a 50 nm cube might require half a billion lines of ROM software, permanently embedded somewhere in its unthinkably miniscule frame. Drexler and the faithful speculate somewhat about how these data might be encoded. Even molecular memory would be too clunky for a working nanobot that was itself molecule-sized. Something else would have to be found.

A large proportion of the nanobot's onboard code would be required just so the nanobot could receive additional instructions from human controllers, reporting its status and the status of the material around it back to real, macroscale people inhabiting our giants' world. This ongoing instructional code would be many times longer and more complicated than the modulating, demodulating, and sensory-processing software embedded in the nanobot. During each minute of molecular assembly, a quantity of new code equal to one-tenth the total amount of onboard software might have to pour in from the outside. The means of transmission are unclear; it might be radio frequency (RF), it might be something else. It would *not* be molecular or ionic, like our own bodies' interior communications. Nature's elegant techniques of molecular signaling, which have evolved over billions of years, are wet nanotech and therefore anathema to the Drexlerians. To them, the universe is just a big machine with little machine-parts, nested in diminishing layers. No place for chemistry there.

Presumably the human controllers of Drexlerian nanobots would sit in control booths, something like Neil Branda's computer-assisted virtual environment. They would think themselves atom-sized as they pursued their excavation and construction work. And they would have their work cut out for them: They would have to control myriads of nanobots simultaneously. Uncounted swarms of the things would have to beaver away endlessly to create something the size of a gnat's eyelash.

In time (Drexlerian theology goes), nanoassemblers would become so advanced that they themselves, with reference to unimaginably huge quantities of onboard memory, would themselves choose how, where, and when to work. At this point, the troops would be their own general. They would operate individually and in cooperation to fulfill their plans—assembling a toaster, or constructing an exaflop computer chip. In the latter case, the nanoassemblers would have achieved the first known example of true artificial intelligence, including the critical component of conscious intelligence called memory. The nanoassemblers would then, by any definition, be a species of living individuals. Humanity would start by making machines and end by making minds. So goes the theology.

It won't happen. The complexity of this whole scenario is beyond comprehension; it out-natures the very nature that it holds in such contempt. Still, this worldview is necessary to Drexlerianism's core beliefs and structure. These are, in essence, that human brilliance and engineering prowess can and must overcome nature, coerce her, and *outwit* her: Bend the old girl to our will. Drex and the boys don't believe in cooperation with the natural world, nor even in its dominance. They seem to require nature's absolute enslavement. This hubris, coupled with the complexity it generates, guarantees the failure of the technoreligion founded by Eric Drexler. The whole movement is just too intricate. It's a Rube Goldberg construct-of-endless-constructs, and it ignores—to its own downfall—that final determinant of all human projects: Murphy's Law.

Murphy's Law has various expressions. The most famous is: "If anything can go wrong, it will." Alternate wordings include "Events maximize chaos" and, my own favorite, "Mother Nature is a bitch." Murphy is the guy that snarls freeways, crashes stock markets, and forever obviates the myriad code lines required for a Star Wars missile defense. The only anodyne to old Murphy's merry pranks is another, more sensible statement: KISS—Keep It Simple, Stupid.

Nature, as it turns out, *does* keep it simple. That's why the best-laid plans of engineers to change nature so often go wrong: We don't think as she does. She won't cooperate because she can't; and she can't because we don't let her. We haven't learned her style. Nature doesn't work by pouring in data from the top down, compelling obedience. There are no Drexlerian homunculi with white hard-hats that strut around a natural construction site. Nature has a different tack. She works with modular units called cellular automata.

In the exhaustive, forty-column index of Eric Drexler's book *Nanosystems*, there are entries for Exoergic, Hagen-Poiseuille Law, London-Eyring-Polanyi-Sato potential, and hundreds of other esoteric terms. Oddly enough, there is no entry for the one concept that gives the molecular-assembler concept its only chance of ever seeing light in any form whatsoever. That concept is the cellular automaton, or CA.

A cellular automaton (pl. cellular automata) is a mathematical abstraction. It's an identical unit that changes state according to the other CAs that border it. This would seem like much advanced mathematics, intrinsically interesting but with no apparent application in the real world, except for one thing. The world, it seems, is based on the CA.

Dr. Stephen Wolfram, who turns forty-three this year, is an English-American genius who made his millions with a popular computer program called Mathematica. It lets the techno-laity harness the power of advanced math without having to spend years swatting up first principles. Mathematica is an interpreter, an interface. It does for math what the consistent, easy-to-use controls of a modern auto do for the Otto-cycle engine: It puts Mr. and Mrs. Everyday behind the wheel. Mathematica gives its users math mobility.

After making his fortune on Mathematica, Wolfram could afford to indulge his fantasies. One of these was taking several years off to contemplate what Doug Adams called Life, the Universe, and Everything. As he did so, Wolfram came to a surprising but, to him, inescapable conclusion. The approach of current science and engineering is self-limiting, he realized; it is dead-end.

Since Newton, science and engineering have related things—force, material, time—with equations. As a dean of engineering once told me as an undergraduate, "If you can't correlate it, forget it." X plus Y is zero; $K = MV^2/2$; *this* is the same as *that*. Undeniably, our love of such statements did produce some interesting results, Wolfram concedes. But it is like Newtonian mechanics: useful, but in a highly restricted area.

Newton is an excellent example of what's wrong with modern science. "Newtonian universality" isn't universal at all. It's what mathematicians call a special case. Under certainly highly circumscribed conditions, Newton seems all-knowing. Change those conditions, and his worldview collapses.

Nature and Nature's laws lay hid in night; / God said: "Let Newton be!" and all was light! So wrote Alexander Pope when the eccentric English scientist, self-poisoned by experiments with heavy-metal vapors, was seen throughout Europe as a scientific god. A modern amendment to Pope's epitaph might be: *God said: "Let Newton be!" and some natural processes were less obscure, some of the time, under highly constrained conditions.*

Pope's couplet was rhyming, rhythmic, and memorable; but there's more to science than good writing. Stephen Wolfram thinks the key to a broader comprehension of nature, and by extension to our humbler (and thus more successful) modification of it, is the cellular automaton.

In a 2001 interview with the London magazine *New Scientist*, Wolfram noted that equation-based mathematics "worked really well for Newton and friends, figuring out orbits of planets and things, but it's never really worked with more complicated phenomena in physics, such as fluid turbulence. And in biology it's been pretty hopeless."

Recognize something? This chaotic natural mess is exactly the context in which Drex and the boys soberly propose to unleash a countless throng of nanomanipulators. Here's a safe prediction for you: They won't work. Ever. But there's a chance, an outside one but the only one, that some types of nanobot based on cellular automata *will*. They'll be simpler than the Drexlerian nanomechano critters, but they will be able to perform certain actions that are still beyond the province even of Neil Branda's molecule-manipulating molecules.

CAs are biomimicry carried to its ultimate expression. According to Wolfram, nature herself works via cellular automata: The CA is The Way. It's how things self-assemble and self-organize, two major goals of nanotech. It's how ants, honeybees, and termites spontaneously change from N billion individuals to a single corporate thing, the nest or hive. It's how waves behave, and how the great rogue-wave solitons that sink ocean-going supertankers are generated. CAs also seem to be the way that atoms become molecules, molecules make up cells, and cells turn into living, breathing, thinking entities such as ourselves.

STAGGERING TOWARD BETHLEHEM

A good illustration of top-down inadequacy is the walking problem. For decades, traditional equation-based science and engineering racked their brains to construct a macroscale robot that can walk. Not over rough ground, mind you; nor up and down stairs, turning corners, or stepping off curbs. No, the aim was merely to develop a legged walking mechanism that could get itself down a polished hall or around a clear, barrier-free room. Nothing more—yet traditional engineering found it impossible to do this. Robots representing millions of dollars' worth of parts and person-hours lurched, staggered, and fell down as soon as they were turned on.

The closest that equation-based science could come to an effective walker was in a series of vast, ungainly mechanical insects. These had even vaster resources, located externally, which constantly sensed and processed data on the state of the walking machine and its relation to the environment. Unfortunately, it turned out that as long as you regard walking (and other commonplace biological activity) as necessarily directed by some exterior, controlling, structuring intelligence, you cannot move your machine. You cannot move *any* machine; you cannot move even yourself. We walk and we think; we do not walk *because* we think. They're separate functions, performed independently and in parallel. Einstein himself couldn't consciously figure out how to do as simple a thing as reaching for a coffee cup. *Okay, Al, here we go! Increase ventricular blood throughput five percent, then raise heart rate from 85 to 98. Contract left triceps ten percent per second for $\frac{1}{3}$ second, simultaneously contracting left pronator quadratis . . .*

Doesn't work, does it? In a sense, the failed walkers based on traditional "equation engineering" were projections of the engineers who designed them. They were avatars of those who were convinced that nature needs telling what to do. That's why they didn't work. Trying to achieve any kind of molecular engineering by means of remote control and top-down instructions is like saying "the beatings will continue until morale improves."

There's another way, and it has begun creating machines that can master walking, even up stairs and over obstacles. It dispenses with the needless, bollixing complexity of engineers who first pretend to be omniscient and then try to dictate all-embracing structures to an idiot, plastic nature. In place of this intellectual tyranny, the alternative method creates a CA matrix where individual elements deduce the proper actions on their own hook, right at point of application. This makes the alternate technique both simpler and more effective.

In pure math, the unit elements are the cellular automata. In the natural macrocosm, they can be foot soldiers or citizens; in the microcosm, living cells; in the nanocosm, molecules or atoms. After all, the goal of nanotech is *self-* assembly, right? Not *engineered* assembly, which is standard manufacture. You do not get things to self-assemble unless you first stop telling them what to do. You don't order, because that doesn't work. Instead, you persuade. You let nature decide to do what *you* want.

Enter the CA. CAs replace unlimited top-down instructions with the freedom for the basic troops of nature, the molecules and atoms that are the nanocosm's poor bloody long-suffering infantry, to make their own assessments and decisions over a specified range. Command devolves to the trenches—and by God, this works. Patterns of great complexity, elegance, and efficiency, both abstract systems and material structures, spring up almost at once: They self-assemble. What they self-assemble may be an approach, like how to walk. Equally, self-assembly may manifest itself as a steady buildup of material structures that begins automatically as soon as the materials are exposed to certain critical circumstances in the surrounding environment. The information central to these working CA systems is not doled out from Central Command. It exists throughout the system as a kind of diffuse, distributed, democratic resource. Information may, in other words, be an inherent property of matter.

Try and describe a complex system in the traditional, authority-driven, top-down manner of equation science, and you bog down. Turn your viewpoint around, democratize it, and describe things from the other end of the corridor, and the "insoluble" problems fall apart. They're ridiculously, stupidly simple. *That's* how insects move their legs. Or rather, they don't move their legs—their legs move themselves. (Note that the legs aren't *designed* to move; they're not "designed" in the sense of "intelligently constructed with top-down control," for anything.)

CA math neatly and elegantly describes not only the systemic motion of a natural (or artificial) leg apparatus, but also its shape and assembly—correction, its *self*-assembly. CA math also covers the self-assembly of the fibers that make up the legs, the cells that make up the fibers, and the molecules that make up the cells. There's no intentional plan, no ghost in the machine: just a configuration that over time self-organizes, self-optimizes—and operates brilliantly.

Here's how a basic CA system functions, in pure mathematics. A series of individual units, each called a cell, is given the ability to take on a single attribute such as shape, color, or anything else. The system starts off with a single CA. Then others are added, and change, according to a simple set of preselected rules. A cell is influenced only by its immediate neighbors.

Say your medium is a sheet of paper, and your basic cell is a capital letter with one of two states: X or O. Assembly proceeds in the way we Westerners read and write, starting at top left and moving rightward. When the line reaches the right margin it drops down a space, resets left, and starts another line.

O is each CA's default state: that is, each new cell is an O. To begin with, there's only one exception. When a new cell being laid down touches a cell that is also an O, then it is an X.

Okay, those are our initial boundary conditions: nothing else. What we get is this:

```
O X O X O X
X O X O X O
O X O X O X
```

And so on. Already, even with these simple rules, a self-organized CA system arises. Two distinct cell types self-assemble into a regular pattern that repeats itself indefinitely over a bounded plane surface.

Now look what happens when we change an initial rule. This time the cell will be an X only if it comes after three Os. This pattern immediately self-assembles:

```
O O O X O O
O X O O O X
O O O X O O
O X O O O X
O O O X O O
```

This second pattern is not as regular as the first—note the interrupted rows of Xs—and is structured differently; in nature, it could be another type of crystal.

We could also make a new cell X only if it is immediately preceded by two Os *and* does not touch an X on either top or bottom. This is the result:

```
O O X O O X
O O O X O O
X O O O X O
O X O O O X
```

Again, this creates a different crystalline form.

You can't specify what a cell will be by reference to what comes long after it. Each cell can develop based only on what comes before and immediately after. This is a direct reflection of growth and development in nature. No natural atom, molecule, or cell has the wit to know what's coming down the pike; neither does any CA. In fact, you can't say a simple cellular individual has any wit at all. It "knows" only what's already next to it, and even then only over a limited range. A cell (living or automaton) senses only its immediate surroundings. No cell being laid down on line 99 can phone back to head office and check what went down on line 10. And it can *never* phone in and ask what's planned for future construction. A CA doesn't see the blueprint; it has no brains. It just follows rules—simple, invariant, prespecified rules. CAs could not recognize a big picture if they tripped over one. They're myopic.

Despite the simplicity of most CA boundary conditions, with slight variations on those initial rules a parade of mindless cells can self-assemble into patterns of breathtaking elegance and complexity. Look what happens when we make an X form after two Os, except when there's an X adjacent up, down *or diagonally*. Then we get the following:

```
O O X O O X
O O O O O O
X O O X O O
O O O O O X
O O X O O O
X O O O X O
O O X O O O
X O O O X O
O O X O O O
```

This is a passable semblance of random distribution. It would work perfectly on a self-assembling semiconductor material (say, silicon dioxide) that we wanted to be evenly doped with a 2/9 (i.e., 22.222 percent) impurity such as gallium arsenide. Specifying such manufacturing patterns with equations is cumbersome, if not impossible. CA math does it in two lines.

CA patterns stem inexorably from initial rules, which mathematicians call boundary conditions. Now here's an interesting fact: Traditional mathematics says any pattern should be predictable from its boundary conditions. And in traditional mathematics, that is certainly the case. For CAs, however, things are different. Knowing boundary conditions rarely if ever lets you predict how the assembled CA system will look. Usually, the only way to see how a CA system will turn out is to let it run—and wait. The best scientists in the world must stand back and twiddle their thumbs while the CAs accumulate surprises. We must drop all pretence of omniscience and be prepared, in great humility, to be astonished at what emerges. That's Law One for the nanocosm: "emergent properties."

One conclusion from this is that information in CA boundary conditions is of a higher order than the data found in a traditional equation. A conventional equation seems inelegant, almost brute-force, by comparison. Traditional science says: *Do this! Do that! Obey!* CA technique advises: *Act consistently, and do as you will. And surprise us.*

It is my belief and prediction that the information encoded by DNA is essentially oriented toward cellular automata. This is the "higher order of data" that Tom Theis of IBM felt in his bones must be, well, in his bones.

Not convinced? Consider the shell of the conch, a summary term for several genera of large marine mollusks. This sea creature has for uncounted millions of years produced a home that grows with it. The conch shell is a model of mathematical perfection as well as of spare, minimalist beauty. If you take the unoccupied shell of a dead conch, saw it in half, photograph it, and digitize the photograph, your graphics card will tell you that the midpoints of the successive chambers occupied by the living conch describe a fourth-order Cartesian curve—a kind of steadily loosening spiral. The defining equation is complex enough to daunt most calculus students. Yet the conch, which knows no mathematics, has mastered it entirely. What gives? Philosophy has been breaking its head on this question since there were philosophers.

It turns out that the conch shell can be described with incompressible brevity by considering it as a CA system. Each new chamber (cell) is added to the previous cell by linear progression and is made a fixed ratio—say 15/7—bigger than its predecessor. That's all! The conch has no need of Cartesian plane geometry, or Newton-Leibnizian differential calculus, to self-assemble. None of these equation-based inventions would do it any good. The conch works on simpler rules: It's a CA system. It leaves the math to us bigwig humans and is a happy animal. Perhaps it's time for science, especially nanoscience, to follow the conch.

At deeper levels—for example, the transcription of genomic DNA by messenger RNA, or the reading of mRNA by an intracellular ribosome to synthesize proteins—nature may rest on the CA. *All* nature. Flora and fauna; the actions of rivers and the periodicity of volcanoes; the interaction of parasites and the likelihood of plagues; the origin and spacing of great galaxies. According to Stephen Wolfram, the *entire cosmos* may be a CA system. Asked if he's discovered the CA program that generates the universe, Wolfram shakes his head. But, he adds, he has "found increasing evidence that it exists . . . [and] could be as simple as a few lines of code." In other words, God is not *in* the details: God *is* the details. Big, little, or in between, everything in nature manifests some primal, still-undiscovered CA boundary conditions. If this is true, then all the wild complexities of reality, both living and nonliving, rest on something unspeakably simple. Configuring the vacuum—now *that's* intelligent design.

If this does prove to be the case, then someone saw it before Wolfram. In *The Hitch Hiker's Guide to the Galaxy*, Douglas Adams suggested that the whole earth might be a computer, working out in real time the consequences of its own CA boundary set. The most powerful beings in the galaxy had to wait four billion years for our planetary computer to crank out its ultimate answers. (According to Adams, it got the answer wrong, but that's life.)

Wolfram in the flesh is different than Wolfram gauged from his writings and published interviews. In print, he seems cold, egotistical, and austere. But when he gives a public lecture to a mixed crowd comprising physicists, housewives, and bums wandering in off the street to get warm, he's riveting. I heard him speak to a packed hall on Thanksgiving 2002 without notes or pauses, and he held us rapt for ninety minutes. I came away wondering if I'd seen the Newton of our time. Wolfram himself appears to think so; but if he's right, that's forgivable.

Wolfram defines his core quest as "the search for how complex structures come about." These structures may be galaxies, hurricanes, or living cells; but somehow, spontaneously, they arise. "Does nature have some mysterious secret by which complex things are built?" he asks. "It turns out the answer is no. All it takes is for nature to follow the rules of a simple CA program." The process is unlike traditional planning, where humans define in advance what a finished product shall look like: "Nature has neither the ability nor the need to foresee what it will do," Wolfram says.

Wolfram himself appears awed with the possibilities, still largely unexplored, of the revolution he's begun. The CA concept, he says, is like the first telescope or microscope: Wherever one points it, one finds new worlds and wonders.

Many of these new findings comprise elegant new ways of modeling reality. For example, CA programs have easily and accurately reproduced the complex shapes of tree leaves and snowflakes. More important, Wolfram has derived the key points of Einstein's theories of special and general relativity from elementary CA boundary conditions.

Although Wolfram maintains that CA models describe reality rather than generate it, many natural processes seem best understood as CA generators working not in abstract mathematics, but in concrete material. The overall shape and constituent stuff of mollusk shells—the elegant spiral, the robust nacre—are more than miracles of self-assembly. Both, I believe, represent the automatic working-out of straightforward CA programs.

I realize this with a start at Wolfram's Thanksgiving lecture when he displays some close-up photographs of conch shells. Patterns on the shell surfaces record how the leading edge of the mollusk's growing tissue has progressed during its biological self-assembly. The shapes match other shapes produced by a computer running Wolfram's standard CA instruction sets. More proof that nature is an analog computer.

The mollusk is only one example; Wolfram is equally comfortable extending the CA concept to the foundation of reality. An idea that explains the structure of seashells may also shed light on the structure of the cosmos. At the heart of things, says Wolfram, creation itself may be nothing but space. The ways in which this "cold vacuum" relates to itself give rise to everything—matter, energy, dimensions, ice cream, dinosaurs, and us.

No two ways about it. This man is on to something.

The concept that the cosmos is CA-based has two staggering implications. First, the universe requires no continuous input of intelligent design from some all-knowing external source. Instead, nature mindlessly and continuously plots out the consequences of its own CA boundary conditions. In nature, advanced systems don't need builders any more than they need architects. They plan and build themselves, winging it as they go. One result of this never-ending process is *Homo sapiens*—us.

Second, self-assembly is nature's norm—not just in living systems, but in everything. To plod unthinkingly on, either toward a dead end or toward greatness; that's the CA way. The process is all, and the end doesn't matter. "We build toward nothing," writes the English author John Fowles in *The Aristos*. "We build."

The consequences of the CA concept for nanotechnology are, like CAs themselves, both subtle and profound. I think CAs are the final nails in the Drexlerian coffin. No nanomanipulator can ever be built to Drex-specs. Everywhere you look, the concept violates natural laws—in the molecular world of stiction, orbitals, and Brownian motion, and more important, in the more fundamental world of conceptual systems. Drex-specs are too complex; they demand ungarbled transmission of too many terabits of data over too short a time, with no interruption. Conceptually, Drexlerianism collapses under its own weight. It's too Newtonian.

A CA nanomachine, on the other hand, might be possible. Might, I say. I'm not about to fall into Drexlerian hubris and grandly announce "This Will Occur." But a CA manipulator that's a stand-alone, whose embedded software is only five lines long; that has one on–off switch, one work speed, one sensor, one manipulator function, and one imperative—well, *that* might just work.

Here's a trial recipe. Start with one of Neil Branda's claw molecules, which changes configuration to grasp or release a certain type of atom—carbon, say—when subjected to a known stimulus. Preprogram a legion of these to function as cellular automata, attaching one carbon atom to another in a cubic latticework. Unleash a zillion of the things on a pile of charcoal, and read a novel as your soot steadily transmutes itself to diamond. Catalytic allotrope conversion—now that seems worth a Nobel Prize or two.

ONE OF THE DREXLERIANS' grab bag of rhetorical tricks is to utter sonorous warnings about How This Amazing and Powerful New Technology Must Not Be Misused. I suspect they do this because getting people worried about something makes it seem more real. No one loses sleep over impossibilities. In this case, though, without any moral credit to them, Drex and his holy men have got it right. Part of the power of a CA system is that it's automatic; it works; it carries on marching. If it mediates a chemical reaction, any reaction, it will take it through to completion. With CAs, you set and forget.

Of course, you then have to live with the inevitable, unpredictable consequences. It would not do to forget about any reaction mediated by even the simplest CA nanocatalyst. The process would not be a traditional construction project as much as a form of artificially induced self-assembly. As long as the resultant workpiece had a lower molecular energy level—and even failing that, as long as the necessary endothermic

energy were available—a CA nanocatalyst would continue its preprogrammed work indefinitely. If the preprogramming commands were basic
enough to be universal—for example, "make all carbon bonds covalent"—or if the preprogramming decayed to an equivalent state through
a software glitch, an artificial molecule used as a CA nanocatalyst could
become an unimaginable destroyer. In sufficient quantity it might convert
all the carbon in living systems to diamond, thus killing us with the curse
of Midas. Jewels, jewels everywhere, nor any thing to eat.

This state of affairs would make AIDS seem like a day at the beach.
Nothing living would survive—no tree leaf or grass blade; no cat, dog,
bird, or human; no crops or crows or cows; *nothing*. Sapphire-emerald
earth would have become Diamondtown. It might not happen rapidly at
first, but it would happen inexorably, increasingly, via geometric progression. The blight would spread like glittering spots of leprosy until the
spots converged and nothing but glitter remained.

What springs to mind here is terrorism. You thought a nuke or two
was bad? Try petrifying humanity, indeed all life on the planet, into everlasting gemstone. It wouldn't matter that this would destroy the terrorists
as surely as it did their enemies, making any new start for life forever
impossible. The main aim of a certain kind of criminal—particularly one
who believes this world is a transitory holding station for heaven—is
destruction: the more the better. Down with materialism! Paradise awaits!
Après nous, le déluge! Can you imagine a fitter tool than nanocatalysis to
unleash on unbelievers?

In his 1994 novel *Idoru*, William Gibson foresaw how nanotechnology,
all nanotechnology, would necessarily be lumped under the same category as nuclear weapons as the most dangerous activities on earth. I tend
to agree with him. At the same time I'm a cock-eyed optimist and think a
doomsday scenario will not occur. Physical and conceptual difficulties
that the Drexlerians have not imagined make their idea of the nanoassembler permanently impossible. Hence we run no risk, now or ever, of a
crowd of little nanobuggers running amok and, like artificial viruses, converting the entire planet into copies of themselves—what one scientist
calls the "gray goo scenario."

In the meantime, any CA manipulators that appear by 2012 or so
will be subject to controls and restrictions sufficient to curtail their
misapplication by terrorists or anyone else. Compare the strain of *E. coli*
artificially bred by U.S. biotechnologists and christened in honor of the

American Bicentennial. This strain, called *e1776,* was intentionally crippled. It was made to act as a living laboratory for biotechnological experiments, but only under laboratory conditions so strict that exposure even to room temperature kills it at once. I can foresee similar fetters being smithed into a CA nanocatalyst. Any removal from a predesigned work environment (say, hard vacuum) would trigger a suicide switch in the marvelous little beasties. One can only hope.

SLOW HASTE

Imminent pitfalls for nanotechnology will be strictly commercial. Dr. James Xu, a professor of engineering physics at Brown University in Providence, Rhode Island, puts it this way: "We may hit the economic limit before we hit the scientific limit." In other words, things may be possible in the lab much sooner than they appear as salable products.

As we've seen, market pull is more effective than technology push in cranking out new science and technology. As a corollary, it's extremely expensive to develop technology for which no commercial market yet exists. If the U.S. government had not invested billions of dollars to develop atomic weaponry during WWII, then fission power—producing electricity from the spontaneous decay of artificially purified uranium isotopes—would still be only a textbook possibility. The same goes for the titanium-extrusion technologies required to make modern jet fighters. Even today's ubiquitous microelectronic applications began as a way to shrink electronic components for the U.S. space program. At its outset, silicon was anything but cost-effective. Economies of scale came later.

All new technology costs money to develop, Dr. Xu reminds us, and nanotechnology will be no exception. His own cost guesstimate for the new techniques is that they will require "twice as much" as society has paid to date to develop and introduce state-of-the-art information technology.

Yet in a larger sense, all this simply doesn't matter to the nanocosm. The nano-revolution is well underway, and it has too much momentum to stop. Even the last three years' repeated meltdowns in tech stocks have only slowed our approach to the nanocosm, not arrested it altogether.

And yet, if ever there were a place for sober second thoughts, even third and fourth ones, it is here. No, we don't know what we may find as we venture deeper into the nanocosm; we know only that it will be unutterably miraculous. But whatever is in there, only the most careful planning can set noble and workable goals for the nanotechnology that proceeds from it: goals like health, wealth, security, and human happiness, and inventions such as universal solar energy, foolproof medical diagnostics, and pure water pulled literally from thin air. Only deep forethought can guide us into products and projects that fulfill the nanocosm's vast promise in a decent, humane way.

The moral to bankers, venture capitalists, financial angels, individual investors, and all the rest of us citizen-onlookers is simple: Let's walk carefully, folks. *Festina lente,* as the Romans said: Make haste slowly. Think before you research, develop, market, and finance. Look long and hard before you leap. The nanocosm is a tiny realm, but its exploration and exploitation will transform our macroscale world. Let's ensure that transformation is all to the good.

WAR OF THE WORLDS, PART II

NOW AND THEN, amid all the missed planes and cancelled meetings, life comes up with a bit of timing as neat as a star gymnast's. This occurred in December 2002 when Dr. Michael Crichton's nano-scare novel *Prey* hit the stores just as this book went to press.

Prey is the perfect embodiment of many points that I've made here. It's a wonderfully readable technothriller, a fun diversion for a winter's evening by a fire. For a writer like myself, it's a special treat to see a master storyteller strut his stuff. One flicks between two viewpoints: sheer enjoyment of the tale and clear-eyed analysis of how the telling is accomplished. First-person narration? The good guy survives. Sick baby? That instantly establishes that the author is going for the glands and not the intellect. Major plot devices? All imported from *The Andromeda Strain, Jurassic Park,* or both. Isolated high-tech facility containing dreadful threats! Venturing out in small groups to beard the nano-velociraptor! And my personal favorite, what *Mad* magazine calls "you know who gets killed!" Then, boom! It's back to racing through the pages with breathless speed. Pure, harmless enjoyment.

Consider *Prey* as a sober warning, however, and it falls apart. At core, *Prey* rests totally on the worldview of Eric Drexler. All those jaw-dropping improbabilities, from molecular assemblers to nanobots to Same Only

Smaller, are givens: Drex is swallowed whole. Sci-fi has always done this; it's like the economist in the old joke: He mislays his can opener, so he assumes one. It's a great way to launch a story and a lousy way to make an honest critique. Surely there are enough real threats in today's chancy world without our having to imagine more.

Having assumed his Drexlerian can opener, Dr. Crichton proceeds to refashion it into *The War of the Worlds,* Part II. This updated version comes nearly forty years after Orson Welles's original scared the pants off the Eastern Seaboard. Now it's not Mars attacking but humanity's own hubris—in the form of artificial nanobot swarms not brought to heel.

As Dr. Crichton imagines it, our inability to cage our newest golem is a many-splendored sin. It extends beyond molecular assemblers, beyond even the nanobots those assemblers put together, to the group dynamics of the nanobots *en masse.* Those, we are warned, will self-assemble into an autonomous brain. This artificial intelligence will then self-program, interacting with reality in lightning-fast iterations until it becomes a self-mobile, self-aware, alien intelligence.

And a hostile one at that. As Dr. Crichton imagines his nanobot collective, it will see humans as its natural food source; hence the title, *Prey.* Look in the mirror, folks. What you see is this Drexlerian *uber-nanotech's* midnight snack. Beeee-waaaare. . . . And shut the book, shiver deliciously before your hearth's dying embers, and go off to bed and sweet dreams. It's just another horror story, no more rigorous in its technology than Mary Shelley's *Frankenstein* was 200 years ago. Dr. Crichton's scenario has as much chance of coming true as Dr. F. had of stitching half-decayed corpse parts together into a wandering monster.

Dr. C's assumptions, you see, all scatter like a house of cards. Assume the U.S. Food and Drug Administration would set aside the world's most stringent protocols for any foreign molecule to be introduced into a human patient. Assume molecular assemblers of unspecified design but enormous powers would be produced at all, let alone by bacterial fermentation. Assume a can opener.

Better safe than sorry, I suppose: *festina lente.* And yet books such as *Prey* will almost certainly foster a climate of suspicion and mistrust, paranoia almost, that will slow the concrete benefits of much nanotechnology by months or years. One thinks of the Apollo astronauts, stuck in needless quarantine for weeks after touching moon dust more sterile than any operating room on Earth. Avoid all possible catastrophes, sure:

Pare the odds; nail it down. But at the same time, see clearly what *Prey* is really about.

For in essence, this is sacred literature. All religion needs holy writ, and the Church of St. Drex is no exception. Dr. Crichton's latest harum-scare 'em is nothing more or less than the Drexlerian Apocalypse, an attempt to make a faith's central tenets seem more believable by showing the horrendous consequences of nonbelief. Read and heed and tremble, for the end is near! *Prey* isn't really a novel, you see. It's the Revelation to Mike.

William Illsey Atkinson
March 13, 2003

APPENDIX
DEVELOPMENTS FROM NANOSCIENCE
BASED ON RESEARCH FOR NANOCOSM

2–5 YEARS

❋ Venture capitalists who can perform due diligence in science as well as business (one aim of *Nanocosm*)

❋ Flat-screen displays for computers and entertainment DVD that are as bright and as omnidirectional as the best CRT displays now available, yet as energy-efficient and thin as today's best LCD and supertwist displays

❋ Twenty to fifty percent reduction in energy needed to pump fluids

❋ Car tires that need air only once a year

❋ Science (weight, vibration, spectroscopy) of single atoms

❋ Self-assembly of small electronic parts, based on artificial DNA or guest-host systems

❋ Engineered nanomaterials compete with expensive, complex test machines such as GCs

❋ Nanotechnology specialties based on geography (e.g., for the United States, Europe, Asia, and rest of world)

❋ New artificial semiconductors based on proteins

❋ Instant, zero-fault pregnancy tests

❋ Complete medical diagnostic laboratories on a single computer chip less than one-inch square. Commercialization of nanofluidics

- Cosmetic technology, including color-change lipsticks

- Go-anywhere concentrators that produce drinkable water from air

5–10 YEARS

- Erasable/rewritable paper for programmable books, magazines, and newspapers

- Plasmon-based computer RAM puts 100 movies on one DVD

- Powerful computers that you can wear as clothes or fold into your wallet

- Bulletproof armor based on nano-biomimicry of nacre

- Light, efficient ceramic car engines

- Intelligent hearing aids that duplicate the natural ear's ability to distinguish speakers (the "cocktail party effect")

- Japanese perfect recyclability of dangerous chemicals

- Molecular machines (i.e., molecules *acting as* machines) that turn commercial catalysts on and off like light bulbs, saving billions of dollars yearly in commercial chemical manufacture

- "Sleepyhead" computers that wake up only to take keystrokes. Based on "spintronic nanotech"

- Golf clubs so light and efficient that players once again can start walking the courses

- Drugs, and drug-delivery systems, that turn AIDS and cancer into lower-level, manageable conditions—as juvenile diabetes is today

- Ultra-high-speed supercomputers capable of understanding some of the most basic processes of life, such as protein folding (1,000 times more complex than the human genome)

- "Smart buildings" that self-stabilize during bombings or earthquakes

- Pharmaceuticals tailored to the individual: "One Size Fits One"

- Inexpensive solar power that allows cities to get energy by using roads and building windows as sun-collectors

- Nanoscale computing hardware, including transistors, resistors, capacitors, and long-term memory storage

- Traditional categories for science and technology (e.g. chemistry, metallurgy) start to blur

10–15 YEARS

- True AI, or artificial intelligence, machines pass the Turing Test, so that anyone communicating with them via voice or keyboard cannot tell if he or she is talking with a machine or a human

- Paint-on computer and entertainment video displays

- Hand-held supercomputers using analog/parallel nanoarchitecture

- Guyed structures 20–100 miles high, used for satellite launches and direct communication

- "Maxwell Demons"—tiny, semi-intelligent devices that sort single molecules—enable instant and automatic heating, cooling, and material sorting, at zero energy cost

- Elimination of the need for non-laparoscopic surgery, since bodies can be monitored and repaired almost totally from within

- Long-lasting batteries and strong yet light car-body materials that allow quiet, zero-pollution electric cars to dominate personal transportation in First World nations

- Cosmetic nanotechnology, including permanent hair and tooth restoration

- Automatic manufacturing using self-assembly, "tunable" catalysis, and other nano-techniques

- Cellular-automaton math combines with materials nanotech to create limited-function molecular machines, e.g. for dusting

GLOSSARY OF TERMS AND ABBREVIATIONS

Absolute zero Minus 273 degrees Celsius: the end of all molecular vibration [heat]

Active nanotech[nology] Nanotech that uses working mechanisms

Actuator A device converting electrical energy into straight-line mechanical energy

Adaptive system A system of computer hardware and software that senses and adjusts to changes in its external and internal environments, *e.g.* component defects

Aerosol A solution whose solvent is air

AFM Atomic force microscope

AIST Institute for Advanced Industrial Science and Technology [Tsukuba, Japan]

Algorithm Formal computational rule [*cybernetics*]

Allotrope A distinct configuration of an element on the nanoscale: *e.g.* diamond, graphite, or fullerenes for carbon

Ampere [A] Basic unit of electric current in SI [*q.v.*]

Analog Using smoothly varying physical quantities to represent reality

Analete The subject of chemical or biotechnological analysis

Ångstrom [å] SI length unit denoting one ten-billionth of a meter (0.0000000001 m)

Antibody A natural defence molecule tagging a biological invader for destruction

Apatite A hexagonal crystal of calcium phosphate

Archaebacteria [Extremophiles]   Bacteria tolerant of the high temperatures and pressures found near deep ocean vents

Avatar   A graphic interface for a human computer user or an artificial-intelligence system. Often a face that expresses emotion and speech

Bacteriophage   A virus that attacks only bacteria

Baud   Standard measure of dataflow, originally set [for telegraphy] at one pulse per second and now standardized as one bit per second

Benchtop   First and earliest stage of technical scale-up for a scientific concept

Benzene ring   A circle of six carbon atoms knit with three covalent bonds [*q.v.*]

Biel   The smallest assay module on a biochip [*q.v.*]

Binary   Arithmetic system using only two numbers, 1 and 0. Used in computers with digital circuits that are fully on or fully off, with no intervening states. *See* Analog

Biochip [Biological microchip]   A miniaturized device combining biological diagnostics with fluidics, electronics, or both

Bioinformatics   The science expressing genetic information as numerical data

Bio-nano   Nanotechnology derived from or applicable to biological systems

Biomimicry   The technique of duplicating or adapting natural systems for nanotechnology

Biopharm   A company using biotechnology to develop pharmaceutical products

Boundary conditions   Initial stipulations for a mathematical or physical system, including transformation rules for cellular automata [*q.v.*]

bR   Bacteriorhodopsin, a photon-capturing chemical related to visual pigments

Broad-spectrum   Emitting photons with many different wavelengths

Brownian motion   Microscopic jiggling of small particles in solution, caused by random thermal motion of molecules in the solvent

Buckminsterfullerenes   A family of carbon allotropes that includes the so-called buckyballs and buckytubes

Buckyball   *See* C_{60}

Buckytube   *See* CN

Butyl rubber™   A synthetic rubber-like substance used in auto tires

Byte   Two data bits

C_{60}   A molecule comprising 60 carbon atoms arranged in a regular sphere

CA   Cellular automaton [pl. *cellular automata*]. *See* State, CA

Calcium carbonate Limestone, $CaCO_3$

Calculus A mathematics of small quantities invented independently by Leibniz and Newton in the 17th century, very useful in modeling dynamic effects

Carbon An element having six protons and six electrons, essential to life

Cartesian A system representing algebraic concepts visually on a plane graph

Catalysis The function of a catalyst

Catalyst A substance that permits or enhances a chemical reaction without itself being changed

Catenate [*v.t.*] To interlock one or more ring-shaped molecules non-chemically

Cathode-ray tube A device that accelerates electrons

Celsius SI scale of temperature measurement that sets freezing water as zero degrees and boiling water as 100 degrees at standard [sea-level] air pressure

CN Carbon nanotube

Cocktail-party effect The human brain's ability to isolate one voice in a crowd

Cold vacuum The basic fabric of the Einstein continuum: "Empty space"

Collagen The main protein part of connective tissue in most organisms

Collective cognitive imperative What a culture deems extant, and hence visible

Colloid [Colloidal dispersion] Particles or droplets of one substance uniformly distributed throughout another substance, and suspended not dissolved

Commodity A product all of whose units are interchangeable

Composite A material with two or more chemically distinct substances in close mechanical association

Compression Squeezing or imploding force

Conductor A substance that efficiently transmits heat or electricity

Control A test population kept untreated so that treatment effects can be determined

Corpsicles The cryogenically preserved corpses of persons who died in the hope that future science might resuscitate them [*Larry Niven neologism: facetious*]

Cortex Outer layers of the brain, used for advanced processing such as vision

Covalent bond A strong chemical bond in which adjacent atoms share two electrons

Cryogenic Supercold: at or below the temperature of liquid nitrogen

DARPA Defence Advanced Research Projects Agency [U.S.]

dB Decibel [SI unit of sound intensity]

Deadman   Concrete ground anchor for a cable, usually set into bedrock

Dendrimer   A large synthetic molecule with many branches [Greek *dendros*, "tree"]

Diagnostics   The techniques of determining the causes of illness

Diamond   A carbon allotrope [*q.v.*] with an infinitely extensible 3-D cubic crystal latticework, transparent to IR and visible light; by far the hardest substance known

Diffraction   The ability of a wave to change direction or create visual patterns through interference with itself, other waves, or matter

Diesel   An internal-combustion engine, usually burning liquid hydrocarbons, which attains ignition temperatures without an electric spark via high compression of the air-fuel mixture [Rudolf *Diesel*, 1858-1913]

DNA   Deoxyribonucleic acid, a genetic molecule shaped like a twisted ladder

Dodecahedron   A twelve-sided solid

Dopant   A trace impurity used in doping [*q.v.*]

Doping   Evenly distributing small quantities of a trace impurity throughout a matrix

Drexlerianism   The body of belief accepted by Eric K. Drexler and his followers, including imaginary constructs such as molecular assemblers purported to accrete macroscale goods from individual atoms and indefinitely prolong human life

Dumbot   A postulated "dumb nanobot" to carry out a simple preassigned task

Dura mater   Tough outer membrane enclosing the brain

e1776   Bioengineered strain of the bacterium *E. coli*, used for laboratory R&D

E. coli   A common species of gut bacteria, most genotypes of which are benign

Edman degradation   A process for determining a protein's constituent amino acids

Elastomer   An elastic polymer, sometimes natural but usually synthetic

Electron   A negatively charged subatomic particle, $1/1860^{th}$ the mass of a proton

E-M   Electromagnetic

Entropy   Randomness of a system, sometimes expressed as heat per unit volume

EPFL   École Polytechnique Fédérale de Lausanne [Switzerland]

Equation   Mathematical statement of equivalence, general form $X = Y$

Emergent property A property appearing in a self-organized system of less complex modules and unpredictable from module properties, *e.g.*, intelligence in brain neurons

Experimentalist A scientist who examines nature to uncover new facts

Extremophile *See* Archaebacteria

F1-ATPase A molecular motor in certain bacteria, rotating at c.800 RPM

Fahrenheit Scale of temperature measurement that sets freezing water as 32 degrees and boiling water as 212 degrees at standard (sea-level) air pressure

Femto- Combining prefix indicating one quadrillionth (0.000000000000001)

Festina lente [Latin] "Make haste slowly"

FET Field effect transistor

FinFET Field effect transistor with a nanoscale cooling fin

Flagellum [pl. flagella] A whiplike feature that propels some microfauna

Flop The fundamental operation of a logic gate [*q.v.*]

Fluidics The technology of manipulating small quantities of fluid, *e.g.*, for numerical computation or direct process control

Fluorescence Re-radiation of photons at different wavelengths after a time interval

Fractal Mathematically identical whatever the magnification: *e.g.*, a coastline

FRS(C) Fellow of the Royal Society [of Canada]

FSS Fixed space facility: "The Tower" [*postulated*]

FTIR Fast Fourier-transform infrared [spectroscopy]

Fuel cell A device generating electric power without noise, waste, or pollution

Fujita-Theis Rule "A living system packs minimal information, one of whose central functions is to extract from the environment the additional data it needs but does not itself possess"

Fullerenes *See* Buckminsterfullerenes

Gallium arsenide GaAs, a doping agent used to create semiconductors

Galvanized Having an outer anticorrosion layer of zinc

Gamma ray A photon with extremely short wavelength and extremely high energy

GC Gas chromatography [sensitive chemical-sampling method]

Gene A DNA sequence producing one or more characteristics in a living organism

Genome   An organism's total quantity of DNA

Geodesic   A straight, rigid strut connecting the centers of two close-packed imaginary spheres [*structural engineering*]

Geodesic dome   A contiguous part of a geodesic sphere [*q.v.*]

Geodesic sphere   A structural approximation of a true sphere using geodesics [*q.v.*]

Goldberg, Rube   U.S. cartoonist famous for humorous, absurdly detailed plans

Graphite   A carbon allotrope comprising parallel plates with low stiction [*q.v.*]

GT   Gene therapy, i.e. therapeutically replacing or knocking out defective genes

Guest-host system   A smaller [guest] molecule trapped inside a larger [host] molecule, the most advanced example of nanoscale self-assembly. *See* Paneling

Gutted virus   The shell of a naturally occurring virus whose genome has been removed and replaced with artificially synthesized DNA as a vector for gene therapy

HARP   High-Altitude Research Project [Canada c.1958]

HAT   Human artificial templating [synthetic organ manufacture]

Heat death   Final state of the universe, when all matter exists at one low temperature

Helicase   A natural editing enzyme preserving the integrity of replicated DNA

Hemoglobin   Large, heavy protein molecule, carrying oxygen in red blood cells

Hertz [Hz]   SI unit indicating cycles per second

HIAN   High-interface-area nanostructures

Highway-busu   Interurban bus [*Japanese: English loan word*]

Hole [electron]   *See* Quantum tunnelling

HRTEM   High-resolution transmission electron microscope

Hydrogen [H]   The simplest element, usually comprising one proton, one electron, and no neutrons. Hydrogen having one neutron is *deuterium;* having two, *tritium*

Hydrophilic   Water-binding

Hydrophobic   Water-repelling

IMAX™   Proprietary technology for filming and projecting large-screen images

Imbedded software   Programming inserted into a machine at manufacture

IMS   Infrared microspectroscopy

IP   Intellectual property

Indirect proof Logically proceeding to an absurdity as a disproof of initial premises

Infrared [IR] Photons of longer wavelength and lower energy than visible red light

***In silica* [Latin]** As modeled on a computer; literally "in sand"

In silico Back-formation from *in vitro;* Latinate barbarism for *in silica*

***In situ* [Latin]** In place; on site

Isomers Related molecules comprising the same atoms, assembled in different ways

Jink [v.t.] To change direction suddenly and skillfully [*snowboarders' slang*]

Joule Standard SI unit of energy, equal to one watt-second or 10,000,000 ergs

Keratin A protein giving horn and hair their structural efficiency

Kips Thousands of pounds of force

KISS principle "Keep it simple, stupid"

Klick Kilometer [*slang*]

Knockout experiment A biotechnological technique in which individual genes or gene sequences are disabled one at a time to determine their effects on an organism

Laminar effect An effect of gas flow over a solid surface

Latex A milky plant fluid containing natural elastomers

Life cycle The period over which a product is conceived, designed, manufactured, sold, used, and abandoned

Linear processing Computation done in sequential steps. *See* Parallel processing

Linotype A device for setting hot-metal type using a typewriter keyboard

Liter The basic SI unit of volume, about 1 U.S. quart

Lithography Production of components using illumination, masking, and etching

Lithotriptor Device for pulverizing kidney stones without surgical intervention

Logic gate Hardware device that changes state [flops] according to signal input

LOS Line of sight

Lumper A theoretician [*q.v.*] who connects disparate disciplines. *See* Splitter

Magnetic lens   *See* Quadrupole

Market pull   The search by business for new ideas [*See* Technology Push]

Mathematica   Proprietary software that lets lay persons use advanced mathematics

MatSci   Materials science [*slang*]

Maxwell demon   An imaginary "intelligent gate" that reverses entropy [*q.v.*]

MEMS   Micro-electro-mechanical systems

Mesoscale   *See* Microscale

Metastase [v.i.]   To develop colonies from an initial site [oncology]

MFM   Magnetic force microscope

Micro-   Combining prefix indicating one millionth (0.000001)

Microcosm   *See* Microscale

Micrometer   *See* Micron

Micron   One millionth of a meter [0.000001 m]. Also written *micrometer*

Microscale [Mesoscale, Microcosm]   A dimensional range of 0.1–100 microns

Milli-   Combining prefix indicating one thousandth [0.001]

Millisecond   One thousandth of a second

MIT   Massachusetts Institute of Technology [U.S.]

MITI   Ministry of International Trade and Industry, Japan [*Defunct April 2001*]

METI   Ministry of Economy, Trade and Industry, Japan [*Established April 2001*]

Molecule   A chemically bonded assembly or two or more atoms

Molecular assembler   Proposed nanobot of ultrahigh complexity, imagined by Eric Drexler *et al.*, which would build macroscale objects by accreting atoms one at a time

Monocoque   Structural design in which an object's skin bears mechanical load

Moore's Law   "CPU capacity per dollar doubles every 18 months"

MRAM   Magnetic random-access [rewritable] computer memory

mRNA   Messenger RNA [*q.v.*] that conveys genetic instructions to cellular factories

***Multum in parva* [Latin]**   "Much in little"

NaCl   Sodium chloride, a cubic crystal used as a spice [table salt]

Nacre   The tough shell of a mollusk; "mother-of-pearl"

Nano-   Combining prefix indicating one billionth [0.000000001]

Nanoassembler   *See* Molecular assembler

Nano-bio [Nanobiosystemics]   *See* Bio-nano

Nanobooster [Booster]   A person who imputes miraculous abilities to nano-technology while disparaging mainstream technology and science

Nanobot   A nanoscale machine; a dumbot [*q.v.*]

Nanocatalyst   A catalyst produced by, and for use in, nanotechnology

Nanochannel   A nanoscale tube for fluid transmission in fluidics [*q.v.*]

Nanocomposite   A composite whose constituent materials associate on the nanoscale

Nanometer [nm]   One billionth of a meter; one millionth of a millimeter

Nanopore   Having openings [pores] in the approximate range of 1–50 nm

Nanoscale   A dimensional range of 0.1–100 nanometers

Nanosci   Nanoscience [*slang*]

Nanoscience   Experiment and theory about nature on the nanoscale

Nanosistor   A nanoscale transistor

Nanotech   Nanotechnology [*slang*]

Nanotechnology   Consistent, predictable manipulation of nature on the nanoscale

Nanotube   A monomolecular, homogeneous, nanoscale cylinder of atoms: *e.g.* silica or carbon

Nem   One nanometer, 0.000000001m [*slang*]

Neuron   A nerve cell, the unit module of the brain

Neutron   A massive neutral particle within an atomic nucleus

nM   Nanomanipulator [University of North Carolina]

NNI   National Nanotechnology Initiative [U.S.]

NoCal   Northern California [*slang*]

NOx   Nitrogen oxide [generic]

Nucleotide   "Ladder-rung" on a DNA double-helix, one of four types: A, C, G, or T

Oilco   Oil company [*slang*]

Oncologist   A scientist who studies tumors

Optoelectronics   Cybernetics using both electricity and light

Orbital   The shape taken by electrons around an atomic nucleus

Order of magnitude   One factor of ten [10X], used in scientific calculations

Organic   Based on carbon, as chemistry or a molecule [*obsolete term*]

Oscillator   A device that generates regular mechanical vibration

Oxidation   The chemical process whereby one atom [the oxidizer] captures one or more electrons from the outer electron shell of an adjacent atom [the oxidant]

Paneling   Using large, slab-like molecules in nanoscale self-assembly [*q.v.*]

Parallel processing   Computation whose steps are performed simultaneously

PARC   Palo Alto Research Center [Xerox Corporation]

Passive nanotech[nology]   Nanotech involving static materials, not mechanisms

PCR   Polymerase chain reaction, a method for multiplying tiny amounts of DNA

Peta-   Combining prefix indicating 15 orders of magnitude [1,000,000,000,000,000]

Phoneme   A distinct unit of sound in spoken language

Photoelectric effect   The basic function of a solar cell

Photon   The elementary quantum of light and carrier particle for the E-M force

Photosynthesis   A natural photoelectric process, usually performed by chlorophyll

Pico-   Combining prefix indicating one trillionth (0.000000000001)

Piezoelectric effect   The ability of substances such as quartz to convert electrical energy to mechanical energy and *vice versa*

Pixel   A picture element module in a video display

Photocatalysis   The use of light to trigger an irreversible chemical reaction

Photochromism   Reversible chemical change due to the presence or absence of light

Pilatory   Medication or other agent that fosters the growth or regrowth of hair

Pixie dust   A one-atom thick layer of rhenium, applied to a computer hard disk to maximize memory storage capacity and minimize retrieval time

Plasmon   Photon trapped on a metal surface as a standing wave, used in super-RENS data-storage technology [*q.v.*]

Plastic deformation   Restricted material flow, creating high strength at the nanoscale

Polymer   A large, heavy, complex molecule, usually with repeating modules

Polypeptide   A structurally strong natural polymer whose module is a peptide

Pressure   A physical quantity expressed as force per bearing area

Pronator quadratis [Latin; anat.]   A small muscle that rotates the human wrist

Protein   A natural polymer comprising a kinked string of amino acids

Proteomics   The science of protein-genome interaction [*protein* + gen*omics*]

Proton   The massive, positively charged particle in an atomic nucleus

Pseudopod   An extrusion extended by bivalves for propulsion [Latin "false foot"]

Qdot   [Semiconductor nanocrystal]   A molecule-sized bit of matter that re-radiates at characteristic wavelengths when irradiated by visible or infrared light

Quadrupole   A linked assembly of two magnetic bipoles, used as an electromagnetic lens to focus an electron beam in cathode-ray tubes [*q.v.*]

Quantum   An irreducible packet of energy emitted or absorbed at the atomic or subatomic scale (Latin *quantum?* = "How much?")

Quantum tunnelling   Transmission of electrons or "electron holes" (electron absences) without apparent movement through an intervening solid

Rastering   Scanning a surface with repeated passes of a sensor or beam

R-brain   The so-called "reptilian" brain at the top of the human brainstem, responsible for regulating some emotions and autonomous functions

Reductio ad absurdum   *See* Indirect Proof

Reduction   The chemical process whereby a reducing element donates one or more electrons from its outer electron shell [orbital] to another atom

RF   Radio frequency [low-energy E-M wavelengths]

R-factor   Measure of how well [or poorly] a substance or a system transmits heat

RIE   Revolution in everything [*facetious*]

RNA   Ribonucleic acid [*See* mRNA]

Rotaxane   A freely rotating system of two distinct, mechanically bound molecules

Self-assembly   The commonplace ability of natural systems and substances to put themselves together without reference to a top-down plan

Sessile   Fixed; unmoving [*said of life forms*]

Shirotae   "White mysteries" [*Japanese*]

SI   Systeme Internationale des unites: the international standard for scientific measurements, also called the *metric system*

SIG   Special-interest group

Single-domain particle   The smallest naturally occurring permanent magnet

Site-specific mutagenesis   Surgical techniques on precise points of a DNA strand

Slurry   A liquid matrix that carries large amounts of undissolved solids

Snurt   Amorphous anterior member of Andromedan life forms [*facetious*]

SoCal   Southern California [*slang*]

Soft nanotech [Soft lithography]   Production and application of molded surfaces that are smooth at the nanoscale

Soot   An amorphous allotrope of carbon

Solar cell   A material that directly converts incident photons to electric current

SOS   Same only smaller

Space frame   A geodesic flat-panel structure

Special case   Law true under highly restricted conditions: *e.g.* Newtonian mechanics

Spectroscopy [Spectrography]   The study of photons emitted by energetic matter

Spintronics   Technology harnessing the electron's spin, in preference to its charge

Splitter   A theoretician [*q.v.*] who makes increasingly finer distinctions among existing scientific disciplines

Sputtering   Heating a material in a vacuum and depositing the boiled-off vapor on a substrate [*q.v.*]

SQUID   Supercooled quantum interference device, a detector of tiny magnetic fields

Stiction   The mutual attraction of adjacent surfaces, both fixed and in relative motion

SPM   Scanning probe microscope

STM   Scanning tunnelling microscope

SSRC   Smart Structures Research Center (AIST, Tsukuba, Japan)

State [CA]   One of a number of configurations permitted a cellular automaton

Substrate   A matrix beneath a deposited thin-film

Superconductor   A material transmitting electricity with little or no resistance

Super-RENS   Super-resolution near-field structure [*See* Plasmon]

Systems engineering   Discipline uniting hardware and software into a single entity

Tanka   A compressed, lyrical poetic form (*Japanese*)

TCRL   Thalamocortical resonance loop: the brain's 40-millisecond snapshot

Technology push   The impetus for someone originating a new idea to market it

Technology transfer   Moving new ideas to market: the final stage of innovation

Tele-   Combining prefix meaning "at a distance"

TEM   Transmission electron microscope

Terabit   One trillion bits of digitized information

Teramac   Fast, defect-tolerant computing system developed by Hewlett-Packard

Tessellate [v.t.]   To completely cover a plane surface with identical shapes

Theoretician   Scientist who unites observations into laws

Therapeutics   Technologies of alleviating or eliminating illness

Thermionic   Produced by heat [electrons]

Thermoelectric   Converting electricity directly to heat or cooling

Thermoplastic   Readily moldable at higher temperatures

TNF   Tumor necrosis factor, a natural cancer-killing chemical

Topology   Mathematics of physical forms and their transformations

Transistor   Electronic device for modulating input signals

Translation   Sideways mechanical displacement

Tunnelling   *See* Quantum tunnelling

Turing test   Seeing if a computer can converse with humans like another human

TYATS   "The young and the stupid," i.e. bright young workers [*facetious*]

UHRTEM   Ultra-high-resolution transmission electron microscope

Ultraviolet [UV]   Photons of shorter wavelength and higher energy than visible violet light

VDT   Video display terminal

Vermiculite   A silicate mineral derived from mica

Vernier   A finely calibrated device permitting delicate mechanical adjustments

Vestibular apparatus   A small device in the inner ear permitting balance and spatial orientation

Viff (v.t.)   To wave furiously or emotionally, as a tentacle [*facetious*]

Virus   A small genome wrapped in a tough shell and capable of replication only by invading a plant, animal, or fungal cell: "A bit of bad news wrapped up in protein" [*Sir Peter Medawar*]

VLSI   Very large scale integration [computer microchip]

Volt   Basic SI unit of electrical force

Wang Wei   Naturalistic, Buddhist poet of the T'ang Dynasty in China

Watt   The basic SI unit of power, equal to one joule per second

Wavefront   The complete forward edge of a moving wave

Waveguide   A device to intercept and steer E-M waves

Wavicle   Matter behaving like particle, wave, or both, depending on circumstances

Wet nanotech[nology]   *See* Bio-nano

INDEX

ABOUT THE AUTHOR

William Illsey Atkinson has spent thirty years uniting science and literature. He was born in Seattle in 1946, his father a U.S. naval officer and his mother the daughter of a Canadian banker. He grew up in Ontario, Canada, and attended McMaster University there. He is citizen of both the U.S. and Canada, which he compares to having two beloved parents.

After university, Atkinson produced readable descriptions of technical projects for the large steel company STELCO. From 1979 to 1986, he worked as a science writer for the National Research Council of Canada. There, constant interaction with front-line scientists, including Nobel laureates, gave Atkinson unique access to the latest research in biotechnology, chemistry, physics, and engineering.

From 1986 to 1991, Atkinson was manager of communications for Forintek Canada, an R&D institute in Vancouver. In 1991, he incorporated Draaken Science Communications to interpret technology for universities, institutes, and private firms.

In 1997 Atkinson received the Prix d'Excellence in Issues Writing from Dalhousie University for his story on the V-chip, which was developed for use by parents to monitor and control what type of television programs their children are exposed to.

He is also the author of *Prototype*, which reviews some of today's most advanced technology, and a finalist for the 2001 National Business Book Award, the only technology book to be so honored.